FREELANCE D

Other How To Books on jobs and careers

Finding a Job in Computers
How to Apply for a Job
How to Be a Freelance Journalist
How to Be a Freelance Sales Agent
How to Be a Freelance Secretary
How to Become an Au Pair
How to Do Voluntary Work Abroad
How to Find Temporary Work Abroad
How to Get a Job Abroad
How to Get a Job in Hotels & Catering
How to Get a Job in America
How to Get a Job in Australia
How to Get a Job in Europe
How to Get a Job in France
How to Get a Job in Germany
How to Get a Job in Travel & Tourism
How to Know Your Rights at Work
How to Manage Your Career
How to Market Yourself
How to Return to Work
How to Start a New Career
How to Work from Home
How to Work in an Office
How to Work in Retail
How to Work with Dogs
How to Work with Horses
How to Write a CV That Works
Living & Working in China
Surviving Redundancy
Working as a Holiday Rep
Working in Japan
Working on Contract Worldwide
Working on Cruise Ships

Other titles in preparation

The How To Series now contains more than 150 titles in the following categories:

Business Basics
Family Reference
Jobs & Careers
Living & Working Abroad
Student Handbooks
Successful Writing

Please send for a free copy of the latest catalogue for full details
(see back cover for address)

FREELANCE DEE-JAYING

How to become a successful discotheque and radio jock

John Clancy

How To Books

Cartoons by Mike Flanagan.

British Library Cataloguing-in-Publication data
A catalogue record for this book is available from the British Library.

First published in 1996 by How To Books Ltd, Plymbridge House,
Estover Road, Plymouth PL6 7PZ, United Kingdom.
Tel: (01752) 202301. Fax: (01752) 202331.

Note: The material contained in this book is set out in good faith for general guidance and no liability can be accepted for loss or expense incurred as a result of relying in particular circumstances on statements made in the book. The laws and regulations are complex and liable to change, and readers should check the current position with the relevant authorities before making personal arrangements.

Produced for How To Books by Deer Park Productions.
Typeset by Kestrel Data, Exeter.
Printed and bound in Great Britain by
Cromwell Press, Broughton Gifford, Melksham, Wiltshire.

Contents

List of Illustrations

Foreword

The word 'discotheque' was coined by the French to describe bars and cafés which played gramophone records because they couldn't get a band. The term 'disc jockey' is American, and describes the presenter of a radio programme of recorded music. Both have spread all over the world, but have become so much a part of British life that they are now an established institution.

Many people aspire to being a DJ or running a mobile discotheque because they seek glamour, easy money and lots of fun. Few find the first two, but there is plenty of enjoyment to be derived. Radio One's Emperor Rosko wrote a book on the subject in 1976. A decade later, another broadcaster from a disco background, Roy Sheppard, published his views. As technology, legislation and fashion all change so fast, it is time for a fresh tome to guide the aspiring DJ on his or her path.

John Clancy has a wealth of knowledge and experience, and the ability to string the words together on paper to make sense. Although physically challenged by his disability, he has carved out a successful career, and whilst never making the very top, has rubbed shoulders and shared slipmats with those who have. I have been in the business since 1966 and know John well as a member of DJ Associations and a writer for industry publications. Despite my advancing years, I am still learning—you never stop in this game—and I welcome a fresh book about the art and profession of dee-jaying.

Theo Loyla
President, South Eastern Discotheque Association

Preface

Greetings! Welcome to *Freelance Dee-jaying*. This is a book I have written after twenty-five years in what I can only describe as the most exciting industry there is. Working as a DJ in whatever field you choose—and there are many—will evoke feelings of joy and elation, depression and misery, satisfaction and well-being; you'll go through the whole range. If you have the necessary dedication, you'll press on and come out on top. Why do we do it? I think the answer is because when everything comes together on the night, there are no words to describe the elation of seeing a packed dancefloor full of people enjoying themselves thanks to your efforts. The DJ has the power to make or break a successful evening.

Much of my career as a DJ has been devoted to mobile disco an area which can be a positive minefield to the unwary newcomer. There is much to learn and consider when starting up, but this is only one avenue open to anyone wishing to become a DJ. In the ensuing chapters we shall also investigate other areas: working in pubs and clubs, local and community radio, hospital radio, and bedroom mixing. In our ever-changing industry there is always something new to learn; this is what I find so exciting. I therefore hope and trust this book will appeal to, and be of interest to, both newcomer DJs and long-standing veterans.

In his book, *A DJ's Handbook* (Javelin Books, Poole), Roy Sheppard states 'DJing is an art and requires technique; it is not a science, so it is impossible to lay down hard and fast rules.' Likewise, one of the main aims of the South Eastern Discotheque Association is 'To promote the art and profession of the disc jockey.' It is one thing to call yourself a DJ, many do, but it is something completely different to actually be a DJ with a full diary of work.

Being a semi-professional, or even fully professional DJ, in whatever field you have chosen, automatically makes you a self-employed businessman. You will need to quickly master the assorted skills of basic book-keeping, budgetary control, taxation, good accounting practice, salesmanship, self-promotion, and become an electrical technician. And you thought it was all down to purchasing the necessary equipment and getting out there on a Saturday night! How wrong you were, but fear not; by the end of the book all will be clear.

We shall probe the pros and cons of joining DJ associations; consider how best to compile a comprehensive music library; look at licensing and other items of legislation; and investigate compiling a demo tape with which to apply for a job in radio. This is a very grey area amongst hopeful wannabes, and is something which took me many years to discover. Having found out, I'm happy to share this 'secret' with you.

Finally I would like to thank all those who have helped, aided and abetted me in my career. There are so many that I fear omitting some, so I'll just say an enormous 'thank you' to Peter Snell of Peter J. Snell Enterprises for starting me on this crazy career; it was he who got me started doing mobile disco, then later commentary/announcing work at fêtes, sports meetings and so forth, and eventually cable TV.

It is said that behind any successful man there is usually a woman, and I'm no exception. Being a DJ is a somewhat self-centred sort of job, so it gives me the greatest pleasure to at last be able to publicly thank my wife Pat, for taking second place to my mistress, disco, for so long. Being a DJ is a job, albeit a part-time one, which can very easily split a marriage or relationship. You need the support of the right partner, and I'm so very grateful I have.

Still with me? Good, read on.

John Clancy

IS THIS YOU?

Outgoing

Friendly

Confident

Well organised

Adaptable

Responsible

Mature attitude

Positive outlook

Self-disciplined

Tolerant

Articulate

Reliable

Open-minded

Knowledgeable about music

Clearly spoken

Good personality

Smart appearance

Effective communicator

Jack of all trades

Effective telephone manner

Calm in a crisis

Courteous

Punctual

Alert

Respectful

Sociable

Good host

1
Getting Started in Mobile Disco

LOOKING AT YOUR POTENTIAL

The reasons why many people wish to become a DJ and run a mobile disco are as varied as the pebbles on the beach. For some it is simply a fun way to spend a Saturday evening, for others it is a way of earning extra pocket money, and others view it as a stepping stone to bigger and better things. Whatever your reasons for wanting to become one of the many thousands of successful DJs who operate throughout the UK, and indeed the world, you must remember that you'll need more than a modicum of talent, and a genuine feel for the business.

It will cost you a few thousand pounds to set yourself up with suitable equipment and transport, then a lengthy period of waiting to fill your diary with work. It certainly will not all come together overnight, so the first question you must ask yourself is:

- Do I really want to become a disc jockey?

If the answer is an emphatic 'YES', you must next ask yourself:

- Do I have the dedication to stick at it until it all comes together?

Have you got what it takes?

Your reason for wanting to become a DJ might be because having seen a mobile disco working at a family party you thought you could do as well, if not better, yourself. However, behind the glitz and glamour of what you see on stage lies a lot of hard work. The dance floor may well be packed full of happy people, but this is due to the skill of the DJ in selecting the right music to suit the audience. Play the wrong tune and you'll soon find all your dancers will sit down. Do you still think you can do as well, if not better? Good, then read on.

ESTABLISHING YOURSELF

Having decided to go ahead and set up a mobile disco, either alone or in partnership with one or more friends, your first question might be:

- Where do I start?

My grandfather once told me 'If you think twice before you act once, then you'll act twice the better for it,' a shrewd quote which I have always held in high regard. Before you go charging off to your nearest disco equipment retailer to make your initial investment, think long and hard about the direction in which you intend to go.

- Talk to existing DJs and pick their brains.
- Get some of the industry magazines and study them from cover to cover.
- If there is a **DJ association** in your area, go along to a few meetings. They will be pleased to see you. (More information in chapter 12).
- Talk to as many people as you can and read as much on the subject as possible.

Eventually, out of all the dross, some clear facts and figures will emerge.

ASSESSING YOUR VIABILITY

By now you should be starting to get a rough idea of how your new mobile disco will shape up. You will have decided on your music format—**vinyl records, CD**, or a combination of both—and certain brand names of equipment will be starting to stick in your mind.

Knowing your competitors

The next step to consider is your viability.

- Is there much local competition, and if so, how good or bad is it?
- Are you going to be able to offer something they cannot?
- What's the market like in your locality, and how far are you prepared to stretch your territory to get work?

You are in with a chance of succeeding if you've done the preliminary research. At last it is time to equip yourself and take steps to start filling your diary with work.

SELECTING YOUR EQUIPMENT

One of the people to whom you have spoken will be your local disco equipment retailer. You'll appreciate by now that there's a wide, and sometimes bewildering, array of equipment to choose from. Panic not.

The music and its reproduction

You will have already decided whether you are going to use records, CDs, or a combination of both. Traditional vinyl records are slowly being phased out, but you may have a sizeable collection you intend using.

Your first decision will be to what extent you will be using records. If it is only occasionally, then you could purchase just one **record deck** to connect to your **sound mixer**. Do not be tempted to waste your money on an inexpensive hi-fi model. They are not robust enough for our use. The models to look for are Technics, either SL1200 or SL1210, or LAD.

Get it fitted into a **flightcase**, a sturdy wooden box with aluminium edges and corners for maximum protection. You will need to fit a **cartridge** and **stylus**; such **turntables** do not come with them ready fitted. Two suggested brands which are industry standards are Stanton 500 or 680, and Ortofon. This item could cost about £400 complete.

Fig. 1. Typical CD players.
Top: Denon DN-2000F.
Middle: Vestax CD-11.
Bottom: Pioneer CDJ 500.

Recording onto minidiscs

An alternative to using the actual records is to record them on to **minidiscs**, which are similar in many ways to a computer floppy disc. The machine both plays and records the discs, giving a sound quality comparable to that of CD, but it does of course depend on the quality of what you record. If your records are badly worn and scratched, then nothing will improve them. You could expect to pay in the region of £400 for a Sony minidisc player.

DAT

Your third option for incorporating records into your show is to record them on to **DAT tapes**. DAT, or **digital audio tapes**, is a system widely used in the professional recording business. The tape is small, resembling a dictaphone cassette, and can be recorded on one side only. Like the minidisc, they produce an almost CD-like quality of sound, and it is possible to randomly select any individual track. The average price of the machine is much the same as the minidisc.

Cassette

If you intend to use records very occasionally, perhaps because there are certain tracks you do not yet have on CD, then an inexpensive way is to puchase a standard **cassette player**, and a quantity of C10/C15 **cassettes** from your local computer shop or a radio equipment retailer. Put one track on each side, and hey presto!

Having a cassette machine with you is useful as more people are likely to bring you a cassette to play at a gig than a minidisc or DAT.

CD players

Your music will mostly be on CDs, so the first major item you will need is a reliable **double CD player**.

The Denon 2000F has emerged a clear industry favourite. It will set you back around £800: expensive, but you'll find it well worth the money. It is robust and the discs will rarely jump. It incorporates **varispeed** for accurate mixing, and will do most of what you expect from it. There are cheaper models available, mostly based on the Denon design, and some DJs even use hi-fi models, but as with most things you get what you pay for.

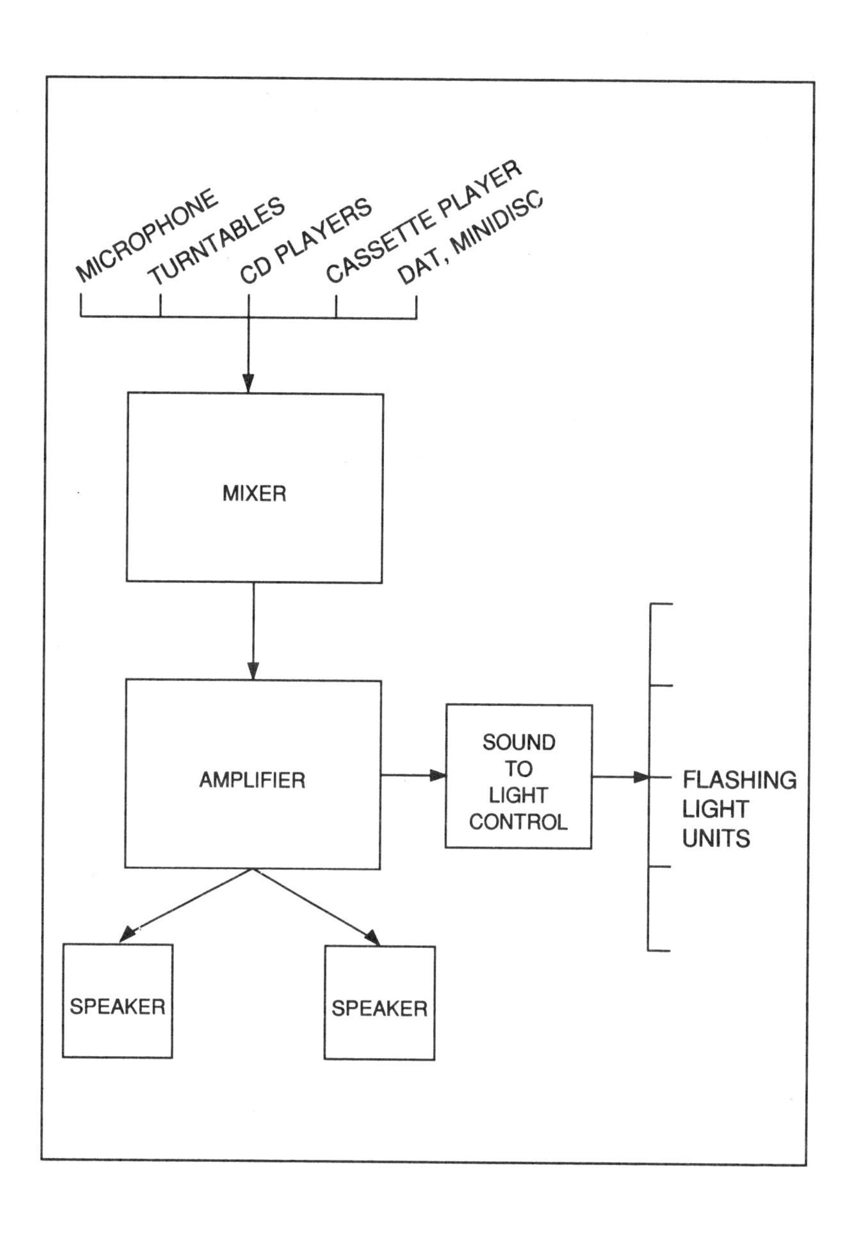

Fig. 2. Typical basic set-up.

Mixers

Your next consideration will be to select a **sound mixer**, a box of electronics with **faders, switches** and **rotary knobs**, into which everything is connected. All these sound sources are mixed together, and the output is controlled by you via the individual faders. You will be confronted by a wide array of models, **four-** or **six-channel**, and with or without **tone controls**.

The more you are able to spend on this, the better the quality. For example, a four-channel model will be cheaper than a six-channel, but it will probably only have one **microphone channel**. The other three are for two CD players and one auxiliary item. A six-channel model will normally give you two microphone channels and two for auxiliary items.

Study what your shop has to offer and carefully consider the options. Citronic is a consistent award winner for its range of mixers costing from around £300.

Microphones

The final item to be connected to your sound mixer is the all-important **microphone**, your vocal link with the audience. Chapter 6 will go into this in more depth and consider how to use it correctly.

Amplifier

Your sound source now needs to be **amplified**. Again the choice is vast, so seek advice. I suggest that to begin with you settle for one of about 400 watts capacity. I have found this to be ample for large and small venues. Many suggest you need amplification in excess of 1,000 watts, but I have never found this to be so. It all depends upon the efficiency of your amp and speakers.

Speakers

Once again you'll need to take advice from your retailer about the most suitable **speakers** for your chosen amp. If your amp is of, say, 400 watt capacity (ie 2 x 200 watts), then each speaker should not be less than 200 watts otherwise the amp could damage it.

Lighting

With your sound system now complete, you next need to look to your **lightshow**. At one time it was the norm to make up a collection of **boxes** and **screens**, and connect them to a **sound-to-**

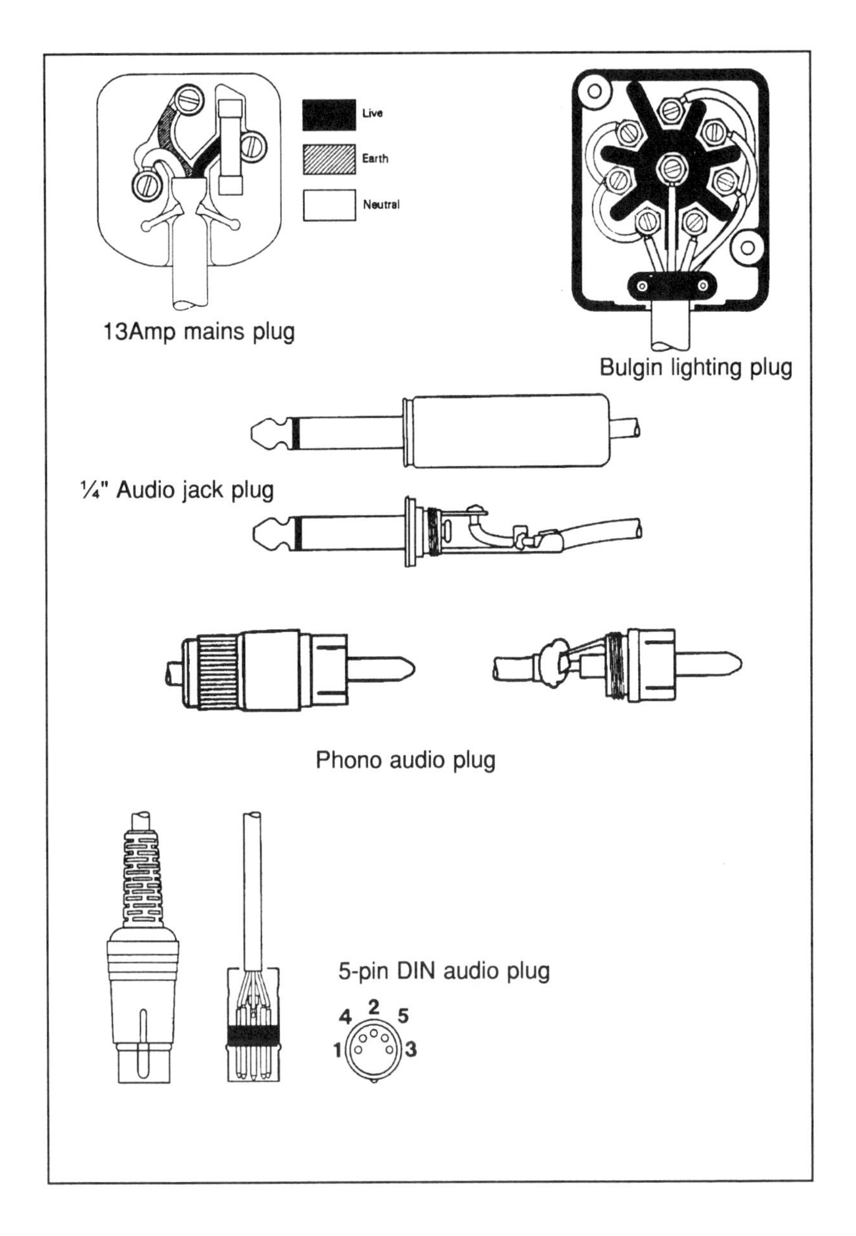

Fig. 3. A range of plugs.

light controller, an electronic box which makes the individual lights flash on and off to the beat of the music. Such homemade boxes and screens have given way to:

- **pinspots**, converted car headlamps which produce sharp beams of light

- **par cans**, floodlights.

Other lighting effects can be purchased which incorporate a built-in sound-to-light controller to control the effect. Remember, the lightshow is something which can be added to as you start to show a profit. It should be:

- flexible,
- stimulating,
- and enhance the disco.

For starters I would suggest that you purchase either four par cans or four pinspots. These will need to be fixed and wired to a **tee bar stand**. This in turn must be connected to a four-channel controller; the three-channel model only requires three lights. Add to this a couple of units such as Abstract Hypermoon, or Martin Ministar or Rainbow, and an **ultraviolet strip light**. One or more centre screens across the front of your show will hide your cables and feet. The screens and par cans or pinspots will be connected to your light controller by cables, on the ends of which are **Bulgin plugs**, the industry standard.

Controllers

Whilst the sound-to-light controller takes care of switching some of your lighting, you will need **switch boxes** to control the rest. If you are unable to make these yourself, your retailer should be able to offer you a ready-made unit. These range from a single box with **rocker** or **toggle switches**, for a few pounds, to ones with **touch-sensitive pads**, computerised and costing several hundred pounds.

PRESENTING YOURSELF

You are now up and running with a pretty good looking and

sounding disco show, but there is just one final consideration before you are let loose on the public. **Presentation**.

You have to decide what image you are going to offer. This will depend on your target market. Teeshirt, jeans and trainers are fine if you are aiming at youth clubs and pubs. If, on the other hand, you are aiming at the more lucrative hotels and corporate functions, then a suit or dinner jacket will be required. If you are working at a social club or family party, shirt and trousers are fine. Save the teeshirt and jeans for loading in and out.

Presentation through chat

To begin with many DJs do not like talking on the mic. Do not worry, it will soon become second nature. Cheeky chat and jokey jibes could become your trademark, and eventually turn into your style of presentation. This is what is termed a **personality DJ**.

Do it with style

Whatever style and method you choose, carry it off with panache and be a star. Watch performers on stage and TV. You won't catch them looking scruffy; casual perhaps, but always neat and tidy. This is what your audience will expect, and should always get.

Presenting your show

Good presentation is vital to the success of any disco show; not only personal presentation, but that of your equipment *and* transport. Some shows look dreadful, with cables strewn all over the stage.

- Ideally all cables should run in a neat, orderly fashion and be stuck down with **gaffa tape**.
- Never mix your **power cables** with your **audio leads**. It could lead to an annoying hum from your speakers.
- A lighting screen makes an ideal centrepiece to your show, and also hides many cables and your feet!
- A piece of black cloth strategically placed does the same job.
- Another option is to use a nameboard across the front, illuminated by a UV strip or spotlight.

Whichever method you choose, remember it is there to hide things but it can also make your show look bigger than it really is.

CHECKLIST

- Check out local competition and forge links, regarding other DJs as colleagues not enemies.

- With careful consideration and advice, select your equipment:
  sound mixer
  CD player, minidisc, DAT or cassette
  microphone
  amplifier
  speakers
  selection of lighting units and switch gear.

- Decide on your style:
  - **Personality DJ**: one who fools around and entertains, with props as well as music.

  - **Mixing DJ**: more suitable in clubs. Mixes the music from one track to the next, maintaining the beat and flow.

  - **Radio DJ**: one who talks and jokes between the tracks or over the 'joins' to establish a rapport with the audience.

CASE STUDIES

Lyn takes up a challenge

Lyn, in her early 20s, is a woman of the 90s and keen to see women make their mark in the world. She realises the world of the DJ is very male-orientated, but certain nightclubs and radio stations make a tiny concession by employing a token lady DJ, which infuriates her.

Lyn decides, with the help of a couple of like-minded chums, to set up an all-girl mobile disco with a view to taking their skills as far as they can and denting this male-orientated bastion.

Gary's enthusiasm gets him going

Gary, aged 16, has long wanted to become a DJ, his enthusiasm having been fired by school discos. As a student he cannot afford the necessary equipment and music.

Gary's father puts him in touch with a company who hire out public address equipment and, sometimes, DJs as well. When they hire out such a combination they also supply the music. This seems like the solution to Gary's problem.

DISCUSSION POINTS

1. Do you have the necessary determination to succeed, having invested so much money?

2. Do you have a clear idea of which brand names you should be looking for in your chosen equipment, following your research?

3. Have you established contacts with other DJs, retailers, DJ associations?

4. Have you established your style of presentation?

5. Conduct a market survey of possible leads for work.

2
Setting Up in Business

GOING INTO PARTNERSHIP

Having made your preliminary investigations into purchasing some equipment, you will have realised that disco equipment is expensive. This is what puts off many hopefuls.

A shrewd move might be to go into **partnership** with someone and share the costs. Such partnerships can work well, in many combinations. Some people might be interested from the technical angle, others might be happy as a roadie or co-DJ. Whether it is your best friend, a relative, or whoever, it is essential to thoroughly discuss a **partnership agreement**. Once all points have been agreed upon a document should be drawn up, signed by all relevant parties, and a copy handed to each.

This might seem excessive, especially if your partner is a relative or friend, but stop and think.

- What would happen if any one partner decided to pull out, leaving the rest with full financial responsibility?

- If the disco were to cease trading, how would any remaining money be split?

A written partnership agreement is essential for all parties concerned.

WHAT'S IN A NAME?

Your next consideration will be to give your disco a name, after which you can get your stationery printed.

JOHN CLANCY DISCOSHOW

An Agreement dated this .... day of .... 19... between .... and

.... hereinafter referred to as the Partners.

In this document, "the Disco" shall mean "the John Clancy Discoshow".

1. The Disco shall be owned and run jointly by the Partners, each owning a 50% share. Each will enjoy 50% of any profits, and bear 50% of any losses.

2. The Disco's trading and postal address shall be ....

3. A bank account shall be opened in the name of the John Clancy Discoshow with the sum of £...., initially deposited, half of which by each Partner. All cheques to be signed by both Partners.

4. Each Partner will agree to place any booking obtained, paid or unpaid, in the Disco's hands. If one partner decides to decline such bookings, for whatever reasons, then the other may undertake that booking privately.

5. Fees for the Disco shall be agreed by both Partners, and quoted in advance of accepting a booking.

6. In the event of the Disco ceasing trading and closing down, all assets shall be sold, and the profits therefrom, together with any money in the bank account, shall be divided equally between the Partners. The bank account will then be closed.

7. During the first two years of this Agreement, each Partner has the right to terminate this Agreement on three months written notice. From the third year onwards, such notice shall be six months.

8. As and when required, either Partner can call a meeting with his co-Partner. Such meeting shall be agreed on two days notice, and each Partner is obliged to attend. At such meetings, a Minutes Book shall be kept to record the purpose and decisions of that meeting.

9. Partners fees shall be 25% of the booking fee.

10. Each Partner shall sign three copies of this Agreement, and shall retain one for his own records. The third copy shall be kept at the Disco's official office for reference at any time.

Signed .... Signed ....

Dated .... Dated ....

Fig. 4. Suggested disco partnership agreement.

- Use your own name, initials, or a combination of yours and your partners' initials, and you need to do nothing more.

- Use a highly inventive name and you'll need to protect it by registering it in accordance with the Business Names Act 1985, otherwise you may find someone stealing or misusing it.

Remember that whatever name you choose may be with you for a long time, so select something which won't sound naff in years to come. Be imaginative and original, and choose a name which will reflect you and the show you have to offer.

MAKING BANKING ARRANGEMENTS

From the moment you put down a sum of money representing your initial investment in the disco, you will need a bank account. This should be solely for your disco business use, and kept separate from your personal account.

Most banks offer a choice of accounts, each with different facilities. I would suggest you opt for a Personal Account, rather than a Business Account, as although both offer a cheque book and full range of banking services, the Business Account will incur some horrendous charges, such as for each transaction you make. Such charges are generally calculated on a quarterly basis and deducted automatically from your account.

Some Personal Accounts pay interest on the balance when in credit. There are no charges made, and you can set up direct debits for your regular payments.

With a bank account in operation, subject to it being kept in credit, you should have no difficulties in obtaining a loan when you wish to purchase new items of equipment or transport.

KEEPING BOOKS AND ACCOUNTS

Being a DJ, playing to a packed dancefloor on a Saturday night, is nothing short of fantastic. However, there is a down side which might not appeal to some but is an equally important aspect of our work: **book-keeping**.

Now that you are a mobile disco, no matter how large or small, and even if you do regard it as only a hobby, you are a small

DATE	NAME	INCOME	OTHER RECEIPTS	DATE	NAME	BILL NO.	EXPEND	EQUIP-MENT	RECORDS & CDs	POST PHONE STATION-ERY	ADVERTS	COMMIS-SION	INSUR-ANCE	ROADIES	VEHICLE EXPENSES	MISC
7.1	Green Dragon Hotel	120.00		2.1	Music Factory	DD	20.00		20.00							
14.1	Pill Mill Club	100.00		4.1	Postage stamps		1.90			1.90						
20.1	Mr Dobson	115.00		8.1	F&W Motor Repairs	1	10.00								10.00	
"	Sale of equipment (projectors)		200.00	14.1	G. Smith		10.00							10.00		
21.1	Miss Jones	150.00		"	Dock Rd Service Station	2	5.00								5.00	
25.1	HM Inspector of Taxes (refund)		50.00	17.1	Woolworths	3	27.00		27.00							
				20.1	Barclaycard	4	150.00	100.00	10.00	10.00	20.00				10.00	
				21.1	B.S. Entertainments	5	22.50					22.50				
				"	S.E. Diso Assoc	6	32.50						12.50		Assoc Fees	20.00

Fig. 5. Suggested layout of cash book.

business. As such you have to abide by the same rules and regulations as others, be it your local corner shop or a multi-national conglomerate. Books of meticulously and carefully kept accounts will have to be kept.

THE TAXMAN COMETH

As you are charging a fee for your services, HM Inspector of Taxes will be seeking to extract his percentage of your income. You will have many legitimate expenses and, subject to your having receipts, they will mostly be tax deductable.

It is in your own best interest to maintain a well-kept accounts book detailing *all* income and expenditure. Even if you do not choose to inform the Tax Inspector of your activities, rest assured he will find out for himself in time. He is a wily bird. If it is years after you started the business, think how difficult it could be to prove your income and expenditure without proper accounts. He will always make a charge on the high side; it is up to you to persuade him to decrease it.

Using accountants

Recent legislation has made it possible for small businesses to conduct their own tax affairs. I should warn you it is a very complex matter, and would strongly advise you to seek the assistance of a qualified chartered accountant. He or she is a vital part of your team.

If your income is low to begin with, you may not have to pay any tax at all. The self-employed are taxed on their profit. No profit, no tax. If your books show a loss you may be entitled to off-set it against your regular PAYE. Determining your profit or loss is of vital importance. To assist your accountant to do the best for you, list every conceivable expense you incur backed up with a receipt. If it is irrelevant, they can always ignore it. Always get a receipt for any item of expense, no matter how small, but if it's not possible use petty cash vouchers.

Paying VAT

Whilst it is compulsory to pay income tax on your income as a DJ, it is hardly likely you'll be required to pay VAT, a tax levied on top of the cost of your service or product. Such tax is collected by HM Customs and Excise, but the figure at which it becomes applicable is sufficiently high to preclude most discos. Your accountant will be able to advise you further.

KEEPING DIARIES

By now work should be starting to trickle in. Are you free? You will need *two* diaries, to be kept identically. One should be on your person at all times, the other at home by the telephone. When you're out, mum/dad/husband/girlfriend, etc, will know exactly your availability should an enquiry come in. A potential customer will not hang around and call back later.

INSURANCE COVER

Your final piece of official paperwork covers insurances. There are three that you will need:

- public liability insurance
- equipment insurance
- vehicle insurance.

Public liability insurance

This covers you should someone sustain an injury or have an accident attributable to you. A well-meaning caretaker or guest at the party offers to help carry out your equipment; they drop it on their foot, and in the post on Monday morning is a writ suing you for damage and incapacity. Where do you stand if uninsured? Membership of a recognised DJ association, as discussed in chapter 12, usually includes PLI cover. It is a condition of membership.

Equipment insurance

This is another vital insurance as disco equipment is keenly sought by the light-fingered brigade. It can either be sold on or used by themselves; either way they can make money at your expense. Favourite times for helping themselves are during your loading or unloading at a gig, when the vehicle is at its most vulnerable. Be watchful. Having paid a few thousand pounds for your equipment it makes sense to spend another few pounds insuring it.

Vehicle insurance

By law a vehicle must be insured for, at the very least, third party.

However, the insurance cover you have on your vehicle generally covers you only as a private person. If you use your vehicle to transport your disco equipment, either within it (as in a van or estate car), or use it to tow a trailer, you must notify your insurer. Otherwise, when you claim on it following an accident you could find your cover to be null and void; although you paid the premiums your cover is worthless and illegal.

Finding an insurer to cover you as a mobile disco is extremely difficult. We are looked upon as irresponsible, drunken louts living in a world of drink and drugs. Insurance surveys have shown us to be heavy drinkers, travel late at night when we are tired, and thus more prone to accidents than most drivers. Do not be upset by this attitude. We are in the same category as publicans, journalists, book makers, sportsmen and even scrap metal dealers! The only advice I can offer is to shop around, but do *not* be tempted to stick to a private person's insurance cover. It is not worth it.

CHECKLIST

- Draw up a partnership agreement in liaison with all interested parties. Get it checked by a solicitor or your accountant, signed by all concerned, and hand each a copy.
- Decide on a suitable name for your disco.
- Open a bank account.
- Understand the importance of good book-keeping practice for tax purposes.
- Understand the importance of the two-diary system.
- Arrange all necessary insurance cover.

CASE STUDIES

Lyn sets up the business

The all-girl disco gets started. Having decided to set up a disco with some friends, Lyn gets to know some local DJs who tell her of the various aspects of setting up. She draws up a partnership agreement, organises the book-keeping and appoints an accountant. The whole team joins the local DJ association where they learn about the problems of getting insurance cover. With the association's help she gets it sorted out.

Gary finds some things are simplified

As Gary is using someone else's equipment, which is transported to and from the site for him, much of the business side of things will not apply. However, he must keep records of all income and expenditure for tax purposes, and use the services of an accountant. He also escapes the penalty of high vehicle insurance premiums through not needing special insurance.

DISCUSSION POINTS

1. What is the significance of a partnership agreement and good accounting practice?

2. Why should emphasis be placed on a disco name?

3. Are your insurances in order, and are you fully covered?

3
Promoting and Publicising Yourself

GETTING YOUR NAME KNOWN

Your mobile disco is now firmly established on a sound business footing. It is time to start actively seeking work. Without promoting and publicising your services no one knows of your existence, and your telephone will remain silent. You will need a promotions package with which to sell your services. Each component needs to be assessed for its role in securing you work.

USING PROMOTIONAL AIDS

Business cards

This will probably be your first point of contact with a potential customer. Make sure you have a plentiful supply on you at all times. When asked for a card never mutter apologies and offer your telephone number scribbled on a beer mat. Keep business cards in your wallet, record/CD cases, diary, in fact anywhere and everywhere.

Your business card says a lot about you, so take time to design it carefully. It can be produced in a variety of finishes—card, plastic, leather, glossy, matt, mirrored, hologram, luminous—the combinations are almost endless. You are seeking to achieve a card which stands out from all others and attracts the eye. Remember, it may be pinned on a noticeboard with many others. Steer clear of those computer-generated ones which come from machines in departmental stores and shopping precincts. They may well be cheap'n'cheerful, but they are bland and say nothing about you—or do they?

Fig. 6. Business cards.

The card needs to include:

- your disco name
- a catchphrase or slogan you might have adopted
- your telephone number. Resist adding '24-hour service' to show you have an answerphone. It is unnesessary.

Show cards

Having made your initial contact with a prospective customer, they will want to know more about you and the services you offer. For this you will need a simple leaflet or brochure, known as a show card. This can range from a single-sided piece of A5 paper or card to A4 printed on all four faces when folded in half. It all depends on how much you wish to say about yourself, and whether you wish to include photos or artwork.

A slightly more expensive method is to have such information printed on the reverse of a photograph of you and/or the disco. A simpler method of producing photocards is to use Letraset to imprint your name and telephone number on the photo, then ask your local photo processing shop for as many copies as you wish. They look extremely smart.

Photographs

In my experience the most difficult promotional item to obtain is a professional-looking photograph of my show. I have used a variety of different cameras, in different settings, but all to no avail. I suggest, if you do not know of a reputable lensman, contact your local library to ask for a contact telephone number for your local camera club. I'm sure from within their ranks you'll find someone willing to undertake the job for a small fee.

Your complete promotion package should contain:

- printed headed notepaper
- business cards
- show cards
- photographs of yourself and/or your disco.

SECURING A FLOW OF WORK

Much of our work comes in by referral. People see us working and

JOHN CLANCY DISCOSHOW

Established 1978

John Clancy & Neil Hannant

It is often said, "A disco is a disco -- a conglomeration of flashing lights, loud pop music, and a disinterested DJ". Whilst this may be so with some mobile discotheques you may have seen, it certainly does not apply to the JOHN CLANCY DISCOSHOW. With an extensive range of equipment, a wide choice of music, and two contrastingly different, experienced DJ's, you can never be quite sure how the show will be presented. For one function we may use compact discs only, whilst at another, conventional records on a standard stereo twin deck unit, and at another, a combination of both. When it comes to music, it is left very much up to you, the client. From our varied library you can choose almost anything from the '50's right up to the '90's. Then of course, the lightshow will reflect the function.

Our "Senior DJ", JOHN CLANCY began his DJ career in 1970 in hospital radio before moving on to BBC Radio Medway, Talking Newspapers for the Blind, commentary/announcing work at fetes and sports meetings, and since 1978, has been Race Commentator for Swale Motor Racing Club at Iwade Raceway. If you have a cabaret night in mind therefore, JOHN is the ideal compere to bring the evening together. He is a member of the South Eastern Discotheque Association and the Thames Valley DJ Association.

Joining John in 1991, NEIL HANNANT, an up-and-coming young DJ who ran his own mobile disco for several years. NEIL is the technical wizard of the team, and should any problems occur on the night, either with the disco or supporting acts, NEIL will be there, screwdriver and soldering iron at the ready! Whilst John can relate to the older clients, NEIL is a firm favourite with the youngsters.

Apart from the foregoing, what else can you expect of the JOHN CLANCY DISCOSHOW? Well, we carry Public Liability insurance, our van is regularly serviced by a professional mechanic and is covered by full RAC service, and we carry a wide range of spares, right down to styli for the turntables. Our proud boast is that no matter what the function, be it in Aunt Fanny's front room or Wembley Stadium, we can cover it (well almost!).

Our sole aim is your satisfaction. Book with confidence.

Printed by Atlantic Print - 0233 624538 -- Exclusive printers for John Clancy Discoshow

Fig. 7. Show card.

ask for a card for future reference. Others might ring and say they'd heard from a friend that you are quite good. We cannot depend upon this for a steady flow of work, though, so like any other business we must advertise. There are three main ways.

- Distributing business cards as widely as possible by pinning them to every available noticeboard and sending them to everyone who may need to book discos.
- Publicity in the local press, which is free.
- Paid advertising in local newspapers and telephone directories.

ADVERTISING

Where to advertise

Only you know which is the best medium in which to advertise locally, but be prepared to think literally and laterally. Put yourself into the shoes of someone who wants to book a disco. Where would they look for a suitable one? Chances are it would be in either *Yellow Pages* or *Thomson's* directories.

You might think the local press to be a good place for your advert, perhaps as a part of a wedding or entertainment feature. Fine, but remember, the local paper is seldom kept for any length of time. It is usually read then thrown away within a week, so unless someone wants a disco when the paper comes out, chances are you'll get little response from your advert. Such **display adverts**, as they are referred to, are expensive, so if you are determined to support your local newspaper by advertising in it you might consider running a short classified advert of three or four lines on a regular basis. It will be far cheaper.

Press publicity

Press publicity is a nice line in advertising and it comes free of charge. However, to gain it you must have an interesting story to tell. All newspapers and magazines, national and local, are keen to hear from you. If you have secured a prestigious job in a nightclub, for example, tell the local newspaper about it. You could even send your story to the national disco-orientated magazines as well. Whenever possible, include a photograph of yourself. Keep your story short and sharp with only the relevant details. Editors are busy people who do not have time to wade through masses of irrelevant dross. You stand more chance of being published if your story is brief and to the point. If more information is needed, and your story has caught the eye, a reporter will soon be on the telephone to talk to you. It really is that simple, but remember the five 'Ws':

- who
- what
- where
- when
- why.

Mail shots

In *Yellow Pages* or *Thomson's* telephone directories you'll find a list of social clubs. Many of these use disco entertainment from time to time, so would like to hear from you. Type out a letter of introduction to each of the social secretaries, asking if he/she would consider using your disco. Mention where you found the address (always useful in assessing whether your advertising campaign has been successful), and enclose your promotion package. Such clubs generally do not pay high fees because of the short hours you would have to work (normally 8–11pm) and the fact that you're likely to get several bookings throughout the year. In your letter quote your normal fee, but add that you are willing to negotiate. Most such clubs have a set fee they pay discos anyway.

USING AGENCIES

In *Yellow Pages* or *Thomson's* directories you'll come across another group of people who would like to hear from you: agencies. Before you even consider working for an agency you have to establish yourself as a bona fide DJ, not just a 'wannabe'. Anyone can purchase disco equipment and call themselves a DJ, but will they turn up on time, or even at all if they get a better offer? It does happen.

Why use agencies?

To a potential customer booking us for the first time, we DJs are an unknown quantity. That's why many people book through an agency, believing they will get a decent show for their money. They gladly pay extra for the reliability and peace of mind it gives them.

In working through an agency you must always remember you have their reputation at stake as well as your own. Mess up a gig and the agent will get the blame. Naturally he'll pass it on to you, but perhaps adding he'll never use you again.

How do I get accepted?

Do not take it for granted that just because you applied to an agency they will use you. They are highly selective and seek only the best for their clients. Acceptance may need to be nurtured over a period of time. Send in monthly **date sheets** detailing your availability, and as and when you get new photos of yourself and/or the disco, or items in the press, send copies.

What are the advantages?

There are many advantages in working through an agency, higher fees and a better class of gig being the most obvious. A good agent has two major priorities: firstly to ensure the customer gets a satisfactory show, secondly to ensure steady work for the artistes on their books.

Working for more than one agency

At this stage of your career work will be coming in from many different sources, so you can quite easily work for several agencies.

However, on a point of ethics, if at a gig someone asks for a card, give them one bearing the telephone number of the person who gave you the booking. If this was a booking received through an agency it is likely the enquirer will demand your own telephone number, hoping the fee will be considerably less than that asked by the agent which includes his percentage. Never weaken. Some agents will send in 'plants' to make sure you abide by this ruling. The worst scenario is, as this is in direct breach of your contract, the agent would be perfectly entitled to instigate legal proceedings against you. Word will get around and your future career will be seriously jeopardised.

What will it cost me to join an agency?

Absolutely nothing. The agent will add her fee, typically fifteen per cent, to the overall booking fee. Sometimes she collects this in advance as a deposit. Other times you'll have to send this sum after the gig. All will be carefully explained in the contract you will receive after accepting the booking over the telephone.

You will not be obliged to accept any bookings offered. The agent will ring to ask if you are free on the date in question, and if you are she'll give you all the details of the booking, including your fee. It is up to you to accept or decline it.

How do I apply?

Write a letter of introduction as you did to the social clubs, but this time include

- details of when you started the disco and why
- a full list of the equipment you use
- your musical policy
- business card
- press cuttings
- photos of you in your various attires, ie teeshirt with disco name, casual wear, suit, dinner jacket.

The agency will accept you in the good faith that you will turn up for a booking on time, so naturally they will expect you to have a reliable vehicle, ideally covered by a breakdown service. Send them proof of your membership.

A clause which appears more and more on contracts these days is one which states your equipment has been checked in accordance with current legislation relating to **Health and Safety at Work**, (more about this in chapter 11). Send the agent a copy of your **test certificate**.

All of these things collectively point to the fact that you are serious in your intentions as a DJ. In time the agencies will increase the amount of work they offer you. You may eventually find that one agent will offer you a **sole agency agreement**, which means they will take full control of all your work, a very prestigious move indeed.

SUMMING UP

- Promotion is a way of life. It is all about the way you treat your clients, audiences and other DJs. This is the single factor which will determine your success or failure.

- Uppermost in your mind at all times must be what it is your customer wants in such things as music, dress sense and presentation. What you personally want is of secondary importance.

CASE STUDIES

Lyn gets the work flowing

Lyn and her all-girl disco have an instant gimmick, but have to be careful that their reasons for this course of action are not misinterpreted.

They decide to use agencies as their main source of work, carefully explaining their stand beforehand. Their business cards and show cards clearly indicate they are an all-girl disco seeking to make a statement for the female cause, and are not in the business for titillation.

Gary shows a little independence

Gary is quite happy working for the hire company, using their equipment and music which is supplemented by his own collection. However, he finds their business cards a little uninspiring. Being company cards they do not mention his name, which he has to scribble on by hand when asked for a card. He could be losing much work to other DJs within the company if the enquirer cannot request him by name. He decides to get his own cards printed with his name on, but with the company's telephone number. He also organises show cards for the hire company to send out with contracts for his jobs.

DISCUSSION POINTS

1. What constitutes a promotion package, and what is its role in your job?

2. What advantages are there to be gained by working through an agency?

3. Devise letters of introduction, show card and business card, and ask friends and relations for their views on them.

4
Compiling Your Music Library

SELECTING THE STOCK

The one thing that both mobile and club DJs have in common is the need for a comprehensive **music library**. Equipment presents no problems; with advice, you can select a suitable sound and lighting system, but will your customers remember you for it? I hardly think so unless your sound system was too loud, or your lightshow blindingly overpowering. You are more likely to be remembered, and re-booked, for the sort of music you played. Unless you advertise the fact that you specialise in rock, rave, techno, or whatever, you'll need to carry a good selection of golden oldies from the 50s to the 80s.

Listen carefully to the IBA golden oldies stations to get an idea of what seems popular. Add to this a selection of the current Top Forty, and you're well on your way. Remember that whatever music you buy for your disco has to appeal to a wide range of people. Listen carefully to each track and decide who, apart from you, will appreciate it. Do you envisage playing it in, say, ten years' time? If so, it will be worth getting.

MIXING WITH PROFESSIONALLY PRODUCED MATERIAL

A very useful addition to your music library would be the **megamix albums** available on subscription from either Music Factory or Discomix Club. These albums are extremely useful, not only to those who as yet have not mastered the art of mixing for themselves, but also to those who use them to increase the range of their repertoire. Not every DJ uses them, so it makes it a wee bit exclusive, and think how useful it is when someone wants several tracks by a certain artiste; you can pop on a megamix by that artiste which the fan has probably not heard before. These mixes are available solely to subscribers, and are rarely released commercially.

Such megamixes include a wide range of material from golden oldies to up-front club orientated tracks, Euro and UK, well-known favourites and the more obscure. The two major suppliers of this material are profiled below.

Music Factory

This company is probably best known for unleashing Jive Bunny on an unsuspecting world. Inaugurated in 1986, Music Factory has members in the UK and Europe. Music Factory claims to adapt itself to members' needs by asking all subscribers to complete a reaction report.

The service is available to bona fide DJs only, and you get your mixes on one CD which sometimes includes a bonus mix. A unique feature is the 'toolkit' section, a selection of drop-ins, samples, dialogue and sound effects which can turn even the most basic CD player into a powerful sampler.

Discomix Club

The longest established megamix service of its kind, Discomix Club was set up in the early 80s by former Radio Luxembourg DJ Tony Prince and his wife Christine. It has members in thirty-two different countries.

Unlike its competitors, Discomix Club tries to tailor its service to individual subscribers' requirements by offering either a double pack of two remixed albums, or a double pack of two megamix albums. Alternatively you could opt for a CD of megamixes and selected remixes.

ADDING SPECIALIST MATERIAL

Your complete music library should include:

- a good selection of golden oldies
- a selection of the current Top Forty which will become tomorrow's golden oldies
- an added dash of the odd megamix or two.

But there's more

For mobile disco DJs there is one further ingredient to add: **ballroom music**. Get a couple of albums by the likes of Victor Sylvester, Sidney Thompson, Brian Anderson, or whoever, and even granny and grandad will swear you're the best disco ever.

Ballroom music

To many DJs the thought of playing ballroom music fills them with abject horror, mainly because it is a music form totally alien to them; but it's not that bad, nor difficult to programme into your usual sets. It is the simplest form of music to use, not requiring mixing like other music forms to maintain the flow of dancing. All that is required is a short, simple announcement along the lines of 'Ladies and gentlemen, take your partners please for a waltz' (or foxtrot, quickstep, or whatever).

The three basics you will need are waltz (the slowest), foxtrot and quickstep (the fastest). Most albums of this music tell you what sort of dance each track is, so it couldn't be simpler. Use them sensibly and intelligently, and very soon you'll find you have earned a reputation which puts you slightly above your peers.

Playing ballroom music

It is not possible to advise on how and when to play such music. No two disco bookings are ever exactly the same. You have to learn to 'feel' your audience. I can only tell you how I test the water.

At a typical wedding reception I begin with the bride and groom's chosen favourite tune, followed by another of similar tempo. After that it's a quick announcement inviting people to take their partners for a waltz, followed by a second. I then change the tempo to a quickstep, and if it works, another, but if not after a suitable announcement I switch to a bouncy golden oldie. Later in the evening, after a set of slowies, I repeat the idea, introducing more waltzes and followed perhaps by Jimmy Shand's Gay Gordons which leads rather neatly into my normal party set. Most reputable record shops should be able to recommend a selection of albums suitable for your use.

Listen carefully to your new purchases, use them wisely, and in time I bet you'll enjoy them as much as your usual pop fare.

OBTAINING FREE PROMOS

It is widely believed that one of the perks of being a DJ is getting most of your records free from the record companies. Whilst this may be true for some, notably prime-time radio presenters, it is far from the truth for the majority of us hard-working DJs.

The trail to the top of the charts for a typical potential hit record, as surmised by a typical record buyer, goes like this. The band/singer cuts the record, it's rushed to the shops and, within a matter of weeks or in some cases days, it appears in the Top Ten. Simple, or is it? What has happened behind the scenes of which the average record buyer is unaware?

How DJs help create hits

First the new record has to be promoted and tried out. It certainly will not chart all by itself. It takes a dedicated and determined band of specially selected DJs to bring it to our attention and convince us to rush out to buy it.

DJs have always been responsible for breaking new acts and getting records into the charts. Dance music heard in your local nightclub has not always enjoyed popularity on the radio, as it does nowadays. Before radio accepted the playing of such music, clubs were the only places where you could hear dance music. The record companies had to send promotional copies of new releases to club DJs to bring them to the attention of the record-buying public. Thankfully radio has now given its blessing to such music being played, and it can be more widely heard.

Bringing new music to our attention is in the hands of either the record company's in-house promotions department, or in the case of smaller labels, one of several independent promotion companies set up specially for this purpose. **Promo records,** sometimes also referred to as **white labels,** are not sent to you as an act of kindness, a gift, or to ease your record-buying budget, but to promote that record. They are sent for you to play at every possible opportunity. The company will expect an initial reaction report from you, and regular feed-back as to how the public is accepting it.

Which DJs qualify for promos?

Record and promotion companies go to great lengths to find the right DJs: those who play a lot of new material and those who are prepared to take risks by dropping in material their audiences have never heard before. You can see why it is generally club DJs who make up the bulk of the mailing lists, because this is where most new dance music records are played. It is more difficult for mobile disco DJs to get on to mailing lists by the very nature of their work at parties, wedding receptions and similar family functions where the norm is golden oldies and favourite tunes. It is virtually impossible for any mobile disco DJ to help break a new record.

The other main factor in determining your worth to record or promotion companies is the size of your audience. A DJ who plays three or four nights per week in an 800-capacity club will be reaching many more potential record buyers than one who does, say, a wedding on Saturday evenings now and again.

How do I qualify?

A typical record promotion company, Euro Solution, explain that to qualify for inclusion on their list you would need to be playing to in excess of 1,000 people per week, and that you can, and do, regularly programme new releases into your sets. Initial reaction reports of both the DJ's and audiences' reactions must be sent in within fourteen days of receipt of the record, together with a copy of the Top Twenty tunes the DJ is playing. This chart should also be faxed to the major industry publications on a weekly basis.

The number of available places on such a mailing list is limited, and Euro Solution normally mail out between 300 and 600 copies of each record. The UK is divided into ten regions which means only between thirty and sixty copies go to each region. With so few copies being sent out it is critical to select the right DJs.

Who else qualifies?

Most promotions are targeted at clubs and regional radio, but those stations must have a large listenership and reach the target audience for a new release. Hospital radio unfortunately does not come within that area as their listeners are often not interested in hearing new tunes, preferring old favourites.

Another angle from which to try for a coveted slot on a mailing list is by writing a regular column in a magazine, newspaper, or club newsletter. Local newspapers tend to have an arrangement with a local shop. If there was one vacancy with two DJs competing for it, both working in similarly sized clubs, but one wrote a regular column which included record reviews, it would go to the DJ who could offer the highest exposure.

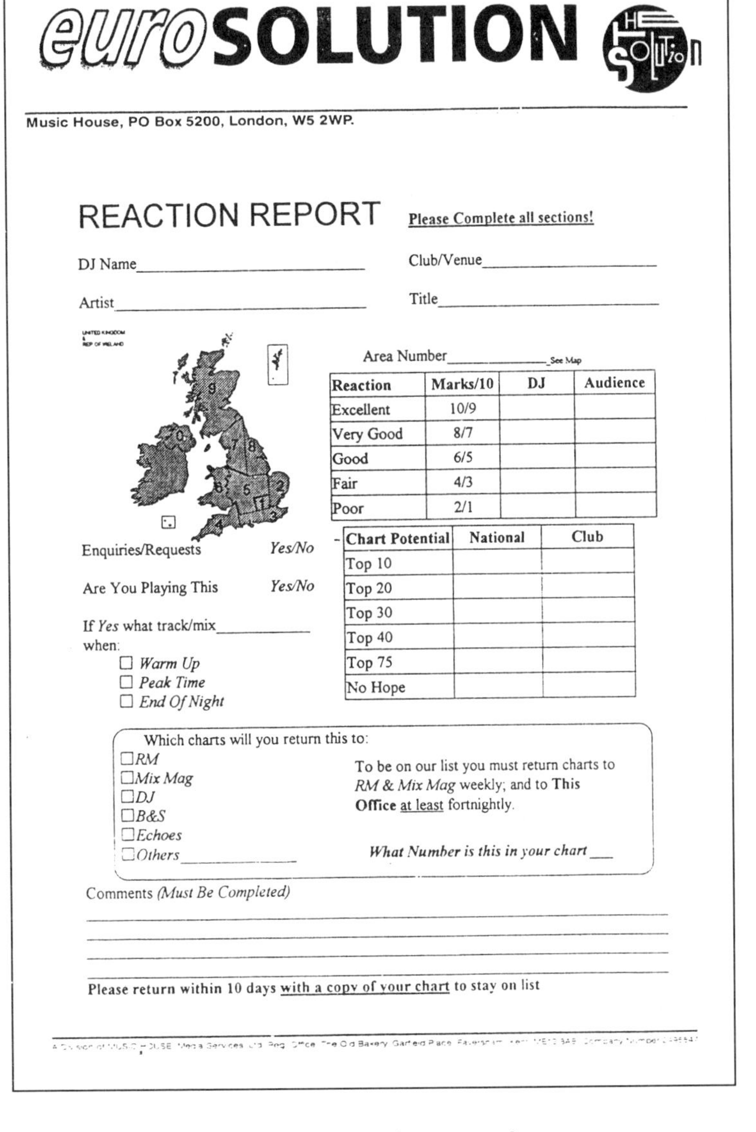

euroSOLUTION

Music House, PO Box 5200, London, W5 2WP.

REACTION REPORT **Please Complete all sections!**

DJ Name__________ Club/Venue__________

Artist__________ Title__________

Area Number__________ See Map

Reaction	Marks/10	DJ	Audience
Excellent	10/9		
Very Good	8/7		
Good	6/5		
Fair	4/3		
Poor	2/1		

Enquiries/Requests *Yes/No*

Are You Playing This *Yes/No*

If *Yes* what track/mix__________

when:

- ☐ *Warm Up*
- ☐ *Peak Time*
- ☐ *End Of Night*

Chart Potential	National	Club
Top 10		
Top 20		
Top 30		
Top 40		
Top 75		
No Hope		

Which charts will you return this to:

- ☐ *RM*
- ☐ *Mix Mag*
- ☐ *DJ*
- ☐ *B&S*
- ☐ *Echoes*
- ☐ *Others*__________

To be on our list you must return charts to *RM & Mix Mag* weekly; and to **This Office** at least fortnightly.

What Number is this in your chart ___

Comments *(Must Be Completed)*

Please return within 10 days with a copy of your chart to stay on list

Fig. 8. Record reaction report form.

Reaction reports

Completion of a reaction report is often a grey area with many DJs, some of whom fear that sending in an adverse report will jeopardise their chances of staying on the list. All the record and promotion companies want is your truthful reaction, good or bad. If you and/or your audiences do not like the record in question it would be helpful to all concerned to know why. Reaction reports do not need to be lengthy documents, most consist of simple tick boxes.

I fit the bill—what next?

If you think you fit these criteria and would like to receive copies of new releases, go ahead and contact a few of the smaller labels to begin with, giving full details of your workload, the venues in which you work, frequency, capacities, and any other relevant information such as radio work, writing, etc. Enclose with your application any relevant press cuttings, flyers, tickets and so forth bearing your name.

Before putting your application package together it would be advisable to telephone first to ascertain that the company has a mailing list, and to whom your letter should be addressed. It may transpire the record label you've selected uses one of the independent promotional companies. Do not be tempted to enlarge on the truth or fabricate any false claims in order to secure a place on the mailing list. Everything will be carefully checked to ensure what you say is correct. Honesty is the best policy.

If at first you don't succeed . . .

Most companies update their mailing list once or twice a year, so if you do not get any response to your letter immediately, do not despair. It may be kept on file for future reference. It's more than likely you'll receive an application form on which to repeat all that was in your letter. It will probably also ask for details of your local record shops and which radio stations you regularly listen to. There are many DJs seeking a coveted place on the mailing list, but there are relatively few places and even fewer vacancies. There is, though, a high turnover of staff in promotion departments, and new bosses have different ideas on which DJs can best use their records.

You do not have to be on any mailing lists. Many top club DJs are not, and many of those who are often do not make full use of the records they are sent, preferring to play the music they have bought themselves.

STORING AND CATALOGUING YOUR COLLECTION

Whether you use conventional vinyl records, CDs, or a combination, your music collection will probably represent the highest running cost you have. Keeping up to date with current chart material can be a financial headache, but when you carefully study the Top Forty chart you will find there is much you can do without. Go for those singles which you feel will still get people up‘n’dancing in, say, ten years’ time. A selection of the rest can be obtained on one of the many compilation albums available.

Do not worry unnecessarily about your audience asking for tracks you do not have. It happens all the time. They can be dealt with in one of three ways:

- Admit to not having it.
- Play a different song by the same artiste, telling your audience this was not the actual song requested, but you felt it was an excellent substitute.
- If all else fails there is the old chestnut of telling them you’ll try to fit it in later. Hopefully by the end of the evening they’ll have forgotten all about it.

Storage

- Having spent so much in building up your music library, it makes sense to protect it as much as possible. Dirty, scratched records and CDs are worthless and unplayable.
- Change the flimsy paper sleeve of the record for a white card one on which you can clearly print, with a felt-tipped pen, the **title, artiste, BPM** and whatever else.
- Keep records in their sleeve when not in use.
- Keep your records and CDs scrupulously clean.
- For CDs you will need a **lens cleaner**, a device which

resembles a conventional CD but which has tiny brushes built into the playing surface. When placed in the drawer it plays as a normal CD, but at the same time the brushes clean the lens.

- Get into the habit of gently wiping each record or CD before playing it.

The playing surface of any record, and more so CDs, should never be touched with your fingers. In time the natural body oils in your fingertips will destroy the playing surface.

Another major cause of irreparable damage to records is a worn **stylus**. You should have a sticker fixed to your consol, noting the date when the styli and/or cartridge were changed.

Carrying cases

Having paid such careful attention to the care of their records and CDs, many DJs then stuff them into beer crates and an assortment of other non-suitable containers, just because they are free. Proper carrying cases are available and not that expensive. They come in a variety of finishes—fibreboard, wood, aluminium—but remember, when they are full they are heavy. Do not be tempted by the cheaper domestic models. They are not robust enough for our use.

Cataloguing

The final thing you need for your music library is a reliable filing and cataloguing system so that any record or CD can be located as quickly as possible. There are many ways in which records and CDs can be filed; only you can decide which works best for you. The most popular method amongst DJs I have spoken to seems to be a system which categorises the music into sections such as rock and roll, reggae, slowies and so forth. Each section is put into alphabetical order of artiste.

CDs need careful consideration as compilation albums contain many different categories. This is how it works for me.

- Take each CD and the booklet out of its plastic case and put into a white card sleeve on which track details are clearly printed or typed. This will save you two-thirds of the total weight and you will get more discs into your carrying case.

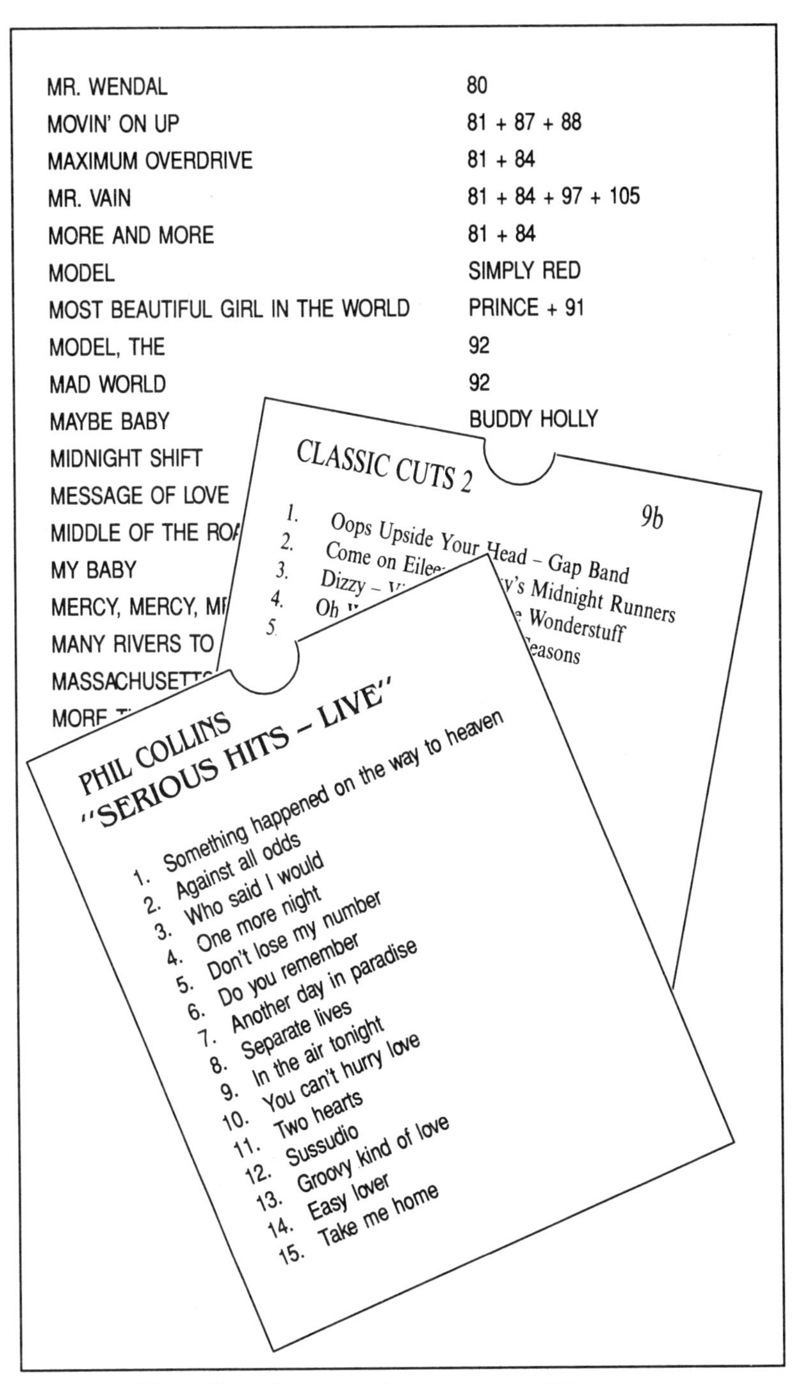

Fig. 9. Page from record catalogue and CD sleeves.

- Each track should be listed in a loose-leaf book in alphabetical order of title. If the CD is a compilation of various artistes, allocate it a number. If the CD is by one single artiste, write that name beside the title.

- Inside the carrying case will be two rows of CDs, one in alphabetical order of artiste and the other in numerical.

- When you look up a specific track in your catalogue it will have either a name or number beside it.

SUMMING UP

- A DJ's music library is a collection of records and/or CDs which reflect the DJ's style of presentation.

- Each collection is personal to the individual DJ, but generally consists of a broad base of golden oldies plus a selection of current Top Forty hits, some megamixes, ballroom music and, if appropriate, new releases.

- All material must be kept clean and stored properly.

- A filing system and catalogue must be kept up-to-date for speedy retrieval.

To get me started can I buy another DJ's records?

Yes, you often see such collections advertised in the local newspaper, but will it contain all that you need? It should contain a lot of useful material, but it depends very much on what sort of DJ you are buying it from. It should, however, prove to be a useful base upon which to start your own collection.

Do record promotion companies offer the choice of records or CDs?

No, by and large they only send out records as it is less costly to produce a limited number of records than CDs.

CASE STUDIES

Gary expands his music library

After a while, Gary finds the hire company's selection of music somewhat limited, and he is never sure what he has in his boxes.

He comes to realise a DJ's music collection is a personal thing and reflects the DJ. He decides to spend some of the money he earns on building up a music collection which will be a truer representation of him than that supplied by the hire company. He will use their material in tandem with his own.

Dave goes Dutch

Dave, aged 27, has gained employment as a DJ on a ferry ship travelling between the UK and the continent. The Dutch owners supply all the music to be played, but it is largely of European origin. He is totally out of his depth, not knowing about this music, so when passengers ask for something specific the best Dave can do is to explain his position, offer up his box and tell people to help themselves. He returns those records he plays to the front of the box, and in time finds all the popular material is at the front and the rest at the back.

Mike and Kath start a mobile disco

Mike and Kath, both aged 35, started their mobile disco as Mike was unable to get a job. Between them they had a sizable record collection, so thought it to be a good way to make some money. They soon come to realize that a DJ's record collection is vastly different from a personal collection such as theirs. They often find they are unable to comply with customers' requests.

DISCUSSION POINTS

1. Devise a suitable filing system for your records and/or CDs. Why does so much importance need to be attached to this?

2. What information needs to be printed on the individual card sleeves, and in what order?

3. How do you put together a package with which to apply for promotional copies of new records, and how does it differ from an application package for work, either at a club or from an agent?

4. What is the basis of a DJ's record collection?

5
Taking on Extra Help

EMPLOYING ROADIES

What would you consider to be the most important part of any successful mobile disco? Some might suggest it is a good, clear sound system. Others might opt for a comprehensive music library. You could even settle for a single item of equipment, or reliable transport. All of these things gel together to make a successful disco, but one essential item which is generally overlooked is a well-trained **roadie**. This is something which cannot be purchased in a shop, but a good roadie is worth his/her weight in gold.

You may already have someone working in this capacity as a part of your partnership agreement. Many DJs do, however, tend to work alone. As a disabled DJ I cannot drive nor hump the equipment myself, so a good reliable roadie is vital to me.

What does the job entail?

A roadie can be either male or female. Sometimes it is the DJ's wife/husband or boy/girlfriend who helps. This is fine all the time they do not distract the DJ from doing their work. I know of lady DJs who have female roadies. Due to equipment becoming more compact with modern technology, roadies no longer need to be muscle-bound hulks. Long gone are the days when we had speakers the size of wardrobes.

It is important he/she can drive. Whilst it may appeal to the DJ to drive to the gig, most would appreciate someone else driving home afterwards. If the DJ has given the gig his all, has helped humping out the gear, I bet he feels like sitting back to unwind on the journey home.

When you take on a new roadie spend as much time as possible before the gig, explaining how everything fits together. Explain how and why equipment, cables and accessories are loaded and unloaded in the precise way they are. Show him/her where fuses are located. It is usually these that blow first, and will be the first thing to check.

DEVELOPING THE ROADIE'S JOB

A good roadie needs a certain amount of technical expertise. If a problem occurs during a gig and a running repair needs to be made, it could be tricky for the DJ to interrupt his performance to attend to it. A good roadie will sense the problem and unobtrusively sort it out.

If the gig is going well the DJ will probably be besieged by customers, all wanting to make requests. The DJ is constantly busy monitoring the music, selecting and cueing up the next track, assessing the audience and thinking about the next direction to move into, so a good roadie should, without being asked, waylay people heading for the DJ and ask if he/she can help.

Roadies as monitors

From time to time the roadie could wander around the venue to monitor the sound quality at the back and sides of the hall. On occasions my roadie has also mentioned customers' comments he has overheard about the disco, which has been useful. It would further be helpful if from time to time the roadie popped outside to monitor how much noise was escaping, possibly upsetting the neighbours.

Prompting the DJ

Occasionally us DJs get a mental block, or an obstinate audience, and no matter what track you put on nothing seems to persuade them to get up and dance. This is what you're there for, and if you fail, even if it's not your fault, you get the blame. On these occasions I turn to my roadie for suggestions. Sometimes he'll remind me of similar gigs and how we overcame that one. He'll also remind me of popular tracks I've forgotten to play; you cannot be expected to remember them all.

Don't overdo it

All discos, large or small, need roadies, but it is important not to overdo it. The larger the disco, the more roadies it may require. If you take too many along, though, the customer will feel you are taking advantage of any hospitality being extended to you. Nothing looks worse than a disco turning up with half-a-dozen or more people. It is not necessary. Even for the largest discos I would suggest two roadies plus the DJ is quite sufficient. Make it clear on your contract how many roadies you will be using.

Further developing the job

Being a roadie, sitting to the side of the stage throughout the gig, could be viewed as boring, so why not get your roadie further involved by also being the **lighting jockey**, or even doing a warm-up spot at the beginning of the evening.

Keep your roadie fully informed

Keep your roadie fully informed about the nature of the gig so that he/she can bring along suitable clothes to change into once the equipment has been set up. Teeshirt (preferably bearing the disco name), jeans and trainers are ideal for humping the equipment in and out, but if you are working at a corporate function or hotel you will be expected to *dress to impress*.

SUMMING UP

- What the minder or bodyguard is to the VIP, the roadie should be to the DJ, constantly anticipating his every need and requirement, leaving him trouble-free to concentrate on the main task of entertaining his audience.

- A good roadie will give the DJ total peace of mind so that he will be able to give his best, without any niggling problems.

CASE STUDIES

Chris finds roadie experience useful

Chris, aged 19, started as a general roadie helping his friend to load in and out at gigs. He was prompted to do this by his love of dance music, and saw it as a cheap night out. He felt too, that it might increase his chances of making new friends.

In time, through watching the DJ, Chris learned how the equipment worked, and eventually asked if he could borrow the consol during the week to perfect his mixing technique. Chris became quite proficient and ended up doing a warm-up spot at the beginning of the evening. News of his skills spread to other DJs who asked him if he would do a guest spot at their under-18s discos.

DISCUSSION POINTS

1. What is the role of the roadie, and how can it be developed within your organisation?

2. Think of other tasks your roadie can take on.

3. Is your show laid out for ease of accessibility should a fault or problem occur?

4. Do you help your roadie by having a fully stocked toolbox of tools, fuses, spare parts and so forth?

6
Addressing Your Audience

Make sure your brain is engaged before you put your mouth into gear. This maxim should be adopted by all DJs, disco and radio alike. When many start on their new career or hobby, the thought of uttering a few words into a **microphone** fills them with terror. It is something they have never had to do before. They soon overcome this fear when asked to make simple announcements such as 'The buffet is now available' or 'Last orders at the bar'.

The microphone is one of the most important pieces of equipment the DJ has. It is your vocal link with your audience. Do not overlook its importance by purchasing a cheap one; this is false economy. It will make you sound unintelligible. If the audience cannot clearly understand you, you may as well not bother. A decent microphone will not cost you a lot of money, and can do much to enhance your voice.

DEVELOPING YOUR VOCAL TECHNIQUE

Even with the best microphone in the world in your hand, you still need to learn the basics of good microphone technique. It is not as simple as holding it in the region of your mouth and shouting your head off.

You must speak clearly, not too quickly, at a regulated pace, and hold the microphone at a distance recommended by the manufacturer. Some need to be held closer to your mouth than others. Whichever model you choose, if you hold it too far away from your mouth you will not be heard and if you hold it too close the bass frequency in your voice will be over-emphasised, causing a popping sound.

- Speaking into a microphone is all about voice projection, not shouting.

Many DJs start in disco work, hoping it will lead to radio. Sometimes it does, sometimes it does not, but either way you will need to develop your microphone technique. There are currently more opportunities than ever before to work in local radio, but the one single thing which should put you in good stead is your voice. Radio listeners like to feel the presenter is talking to *them*, on a one-to-one basis. He/she must therefore have a natural, sincere, friendly tone to their voice. Treat your disco show as a radio programme.

How to train your voice

Few are blessed with a voice which has all the required qualities. You have to work on what you've got and make the most of it. Unnatural accents and insincerity are the biggest turn-off, so get out of that habit quickly. Speak in your normal voice as though you were addressing your best friend, and if your voice is not as deep and resonant as you'd like, try lowering it by a few decibels. It can improve a high-pitched voice. Be in control of your voice at all times by feeling relaxed and comfortable when you speak. This can be achieved by breathing deeply.

Before you make any announcement, no matter how trivial, think carefully what it is you are going to say. Pre-planning eliminates those amateurish 'ums' and 'ers', making you sound much more professional.

CHOOSING THE RIGHT MICROPHONE

When you go to purchase your microphone you will be asked which type you would like. There are two sorts:

- **Omni-directional** which picks up sound from all around and is ideal for use in studios.

- **Uni-directional** which picks up sound from one side only, front or back. This is the one for us disco DJs.

Why a uni-directional microphone?

With this kind of mic the sound of your voice is picked up by the front of the microphone only. This eliminates **feed-back**, that annoying screech caused by sound coming out of your speakers and re-entering the system through your microphone.

Impedance

- There is one more consideration when buying a microphone: **impedance**. You must purchase one whose impedance matches your mixer. Some mixers cater for **high impedance** only, others for **low impedance**, and some for both. Most disco mixers and microphones, however, are of high impedance.
  The differences are:

- High impedance microphones are connected to the mixer by a **single core screened lead**.

- Low impedance microphones need a **twin core screened balanced lead**, which means they can be longer, enabling you to go out on to the dancefloor with it.

RADIO MICROPHONES

Becoming more popular with the DJ of the 90s is the **radio microphone system**. With no trailing leads to worry about you are free to wander wherever you wish, even out of the room. The system has a transmitter built into it, and the signal is picked up by a receiver which is connected to a microphone socket on the mixer.

Once again there are two basic types to consider:

- The **single channel** system, which uses a single receiver. It has one aerial.

- The **diversification** system, which is basically a single channel system but uses two receivers. These two receivers are normally contained in the same enclosure and are configured to provide a single audio output. This system has two aerials.

The difference, generally speaking, is that the diversification system offers a better performance, because it continually monitors the strength of the signal received by both receivers and automatically switches to the stronger of the two.

THE LAW AND LICENSING

The radio microphone system transmits radio signals to a receiver, so it comes within the scope of the **Wireless Telegraphy Act 1949**. However, not all radio microphones require a licence. Wide band radio microphones, with an operating power of 2mW or less, do not need licensing if they operate on the frequencies of 173.8, 174.1, 174.5, 174.8 and 175.0 MHz. They must also conform to the performance specification MPT1345 (or earlier superseded).

If you use a microphone that does not have MPT1345 approval you will require a licence, or risk facing a fine of up to £5,000 and/or six months imprisonment.

Sharing a frequency

When you purchase a MPT1345 approved microphone you must remember it is on a frequency shared by others. It is not exclusive to you, so if you operate in the near vicinity of another user you might well find you are picking up their signal as well.

If you intend using your system in school premises, for example, check first as some aids for the handicapped use narrow band equipment in the same frequency range, and you may interfere with this equipment.

To reduce the likelihood of interference from another user, you might like to consider the new **multi-channel licence** which is now available. It offers five channels in the 190 to 216 MHz range for an annual fee of £130, but remember, if you go for this option you would need to change your radio microphone system completely.

Where to find out more

The Secretary of State has appointed ASP Frequency Management Ltd as agents responsible for distributing licences. If you would like further information they have a leaflet entitled *Radio Microphone Licensing in the UK* (ref. ASP17), available from:

ASP Frequency Management Ltd
Edgcott House
Lawn Hill
Aylesbury
Bucks HP18 0QW
Tel. (01296) 770458.

FURTHER CONSIDERATIONS

Spend time trying out various microphones in the shop, taking a friend or your partner along to get an honest opinion of how you sound. If you are the sort of DJ who holds the microphone in your hand all evening as a psychological crutch, then weight will be another major consideration.

I prefer a **microphone stand**, and use an **anglepoise** as favoured by most radio stations. The beauty of the anglepoise is that you can feed the cable through the arm and keep it out of sight. You will have only about six inches of cable protruding from each end, one to connect to your microphone socket and the other to your microphone.

INTERVIEWING TECHNIQUES

You might think interviewing is more associated with radio than disco. True in part, but you may be asked to announce a guest artiste on stage, or compere a competition. Your audience will want to know a little about whoever you have on stage. Keep it brief and to the point. Couch your questions in such a way as to bring forth a lengthy answer, not a simple 'yes' or 'no'. Never interrupt the interviewee with comments or laughter. Nod silently and let them have their say.

SUMMING UP

Remember the DJs' maxim: never speak for the sake of it.

- Be concise and clear at all times. To achieve this you will need a good quality microphone of an impedance to match that of your mixer.

- If you intend purchasing a radio microphone system, make sure it is marked 'MPT1345 Licence Exempt'.

- You will need to develop your broadcasting voice and practise interviewing techniques.

DISCUSSION POINTS

1. What is the difference between omni- and uni-directional microphones, and what is meant by high and low impedance?

2. Learn to develop and project your voice.

3. Practise interviewing with a friend or partner.

7
Looking After Your Kit

BE PREPARED

It is a short-sighted DJ who goes off to a gig with absolutely no tools or spares, either for equipment or transport. The only time you are likely to develop a fault is in the middle of a gig, so like the Boy Scouts *be prepared.*

I appreciate that many DJs are technically useless. I am myself, I have enough trouble sorting out the three wires in a 13 amp plug! We have to come to terms with the intricacies of mixing music, microphone technique, and audience participation, so we should become acquainted with basic maintenance procedures as well. When an item of equipment develops a fault, it is usually due to a minor problem such as a fuse or cable connection which needs rejointing. I'm not suggesting you attempt to dismantle an item and rebuild it at a gig. If this becomes necessary, you should swap it for a spare which you should be carrying. All equipment should be regularly serviced. Preventive maintenance is always cheaper.

There are two items you should consider carrying, neither of which need cost a lot:

- A spare amplifier of at least 100 watts. It can be secondhand, and mono rather than stereo. If your main amplifier develops a major fault at least you'll have a spare to fall back upon. Without a spare, you have no sound.

- A spare light controller. It need not be as comprehensive as your main one, just something to see you through the evening.

STOCKING UP A BASIC TOOLBOX

To be able to carry out minor repairs it is essential to carry a toolbox containing all the obvious, and the not-so-obvious, items you might need. Keep the box with the rest of your disco gear, and do not be tempted to dip into it when doing a job at home. Chances are you'll forget to put something important back after you've used it.

What should the toolbox contain?

- A definite must is a **soldering iron** and a reel of **self-fluxing solder**. When you use the last of this, replace it immediately
- an assortment of screwdrivers, large and small, ordinary blades as well as Phillips and Pozidrive
- wire strippers
- pliers, both blunt nosed with a cutting edge, and long nosed
- insulating and gaffa tape
- a Stanley or craft knife
- switch cleaner fluid
- spare plug tops, both 13 amp and Bulgin
- an assortment of various fuses
- jack plugs
- drive belts and elastic bands for motorised effects
- a few assorted-sized screws.

In fact, anything and everything that might come in useful.

And the not-so-obvious contents?

- A few elastoplasts
- a blob of Blu-tack
- safety pins (you never know when a zip fastener will fail)
- a needle for digging out that annoying splinter
- perhaps a couple of condoms (very useful as an emergency drive belt if you haven't a suitable sized elastic band and, of course, to waterproof things in the wet)
- a small torch; check the battery frequently.

CHECKING FOR POSSIBLE FAULTS

So there you are, happily fulfilling the role of the successful DJ with a packed dancefloor, when disaster strikes. Your sound system suddenly goes dead. Your initial reaction will be one of total panic. It will not be helped either by the fact that your audience will probably be booing and hurling abuse at you. They can sometimes be very intolerant. This is where a good roadie will come into his or her own.

Going through the signal path

The fault could have come from any one of several sources, so it's important to stay calm and go through the **signal path** systematically.

To begin with, it would appear the **amplifier** is not functioning, or it is not receiving the signal from the mixer. Before you start ripping out leads with a view to changing the amplifier, make sure they are all connected properly. If your amplifier is on the floor, as many are, it is surprising how easy it is to accidentally dislodge a lead with your foot or, as I have done on numerous occasions, brush the **volume control knob** with your foot, decreasing it to zero.

Having checked that all leads are still connected and have no breaks in them, your next move might be to remove the **mixer-amplifier connecting lead** from the consol. Turn up the **amplifier volume control** slightly, and touch the tip of the **jack plug**. If you hear a buzzing sound from the **speakers,** you will know the fault lies within the mixer. It is more likely that a lead needs repairing, or a fuse has blown. Make sure you carry a spare set of leads.

MAINTAINING YOUR EQUIPMENT

All your equipment should be serviced regularly by an authorised and qualified person, and its safety certified under the requirements of the Health and Safety at Work Inspectorate's Portable Appliance Testing legislation, as discussed in chapter 11. There are a number of things you should do regularly to prevent problems:

- Keep all electronic equipment clean and dry.

- Prevent a build-up of dust in your **faders** by squirting **switch cleaner** into them and whipping the **sliders** up and down rapidly.

- **Projectors** generate a lot of heat, which attracts dust into the **fans, vents** and **lens**. Strip them down and clean them often.

- Turntables will run slow if the **bearings** become dry, so periodically remove the **circlip** at the centre of the turntable and lift it from its **bearings**. Lightly oil the top bearing with machine oil. Work the **spindle** up and down until it spins freely.

HANDLING AND PACKING FOR TRANSIT

I find it amazing that problems do not occur more frequently. The equipment is loaded into a vehicle, driven long distances sometimes over bumpy roads, then dragged into a venue and set up. At the end of the evening, whilst most of the items are still hot and therefore very fragile, it is all pulled apart, loaded up and driven home. It is usually at this end of the evening, when you are tired that you are perhaps not as careful as you should be in disconnecting cables and leads. I believe this is where many problems begin.

A periodical check of your leads and cables will not take up much time at home, and will save you embarrassment on the night.

Packing and stacking for transport, be it in a van, trailer or estate car, is of vital importance. If the items are not stacked and packed properly you will have to endure a journey with bangs and crashes each time the vehicle turns. Apart from being a nuisance, think what damage it might be doing to the equipment. Bulbs in particular are very vulnerable.

What can be done to prevent damage?

The proper way to protect everything in transit is by **flightcasing** it. This entails the item being built into a plywood box with aluminium edges and corners. This makes things look very smart and business-like, as well as totally protecting it in transit. It can be done to your consol, amplifiers, lightboxes, in fact, whatever you wish. The retailer from whom you bought your equipment will be able to offer such things, but you can of course make your own. If you do, do not be tempted to use chipboard. It may well be about two-thirds cheaper than ply, but it is a lot heavier and will not last as long.

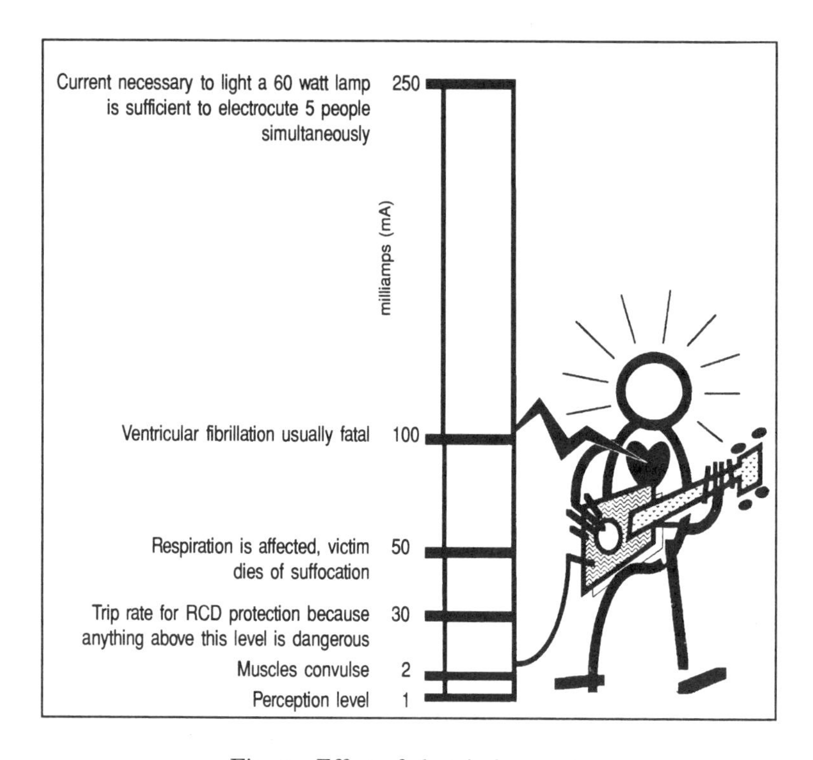

Fig. 10. Effect of electrical current.

For your speakers you can do no better than to purchase a set of **vinyl covers**. Your equipment is now uniformly 'squared' and will fit together more easily when packed.

Finally your leads, cables and smaller items can be stowed into any sturdy box or case. The secondhand shop is a great place to find old suitcases. Carefully coil all leads and cables individually, otherwise they will end up in a spaghetti of a jumble.

AVOIDING ACCIDENTS

Using circuit breakers

Over the years many entertainers have been seriously injured or killed by electric shock whilst performing. Even a small electrical current racing through your body can kill you. It takes only *one-twentieth* of an amp, or 50 mA, to cause pain, paralysis of the

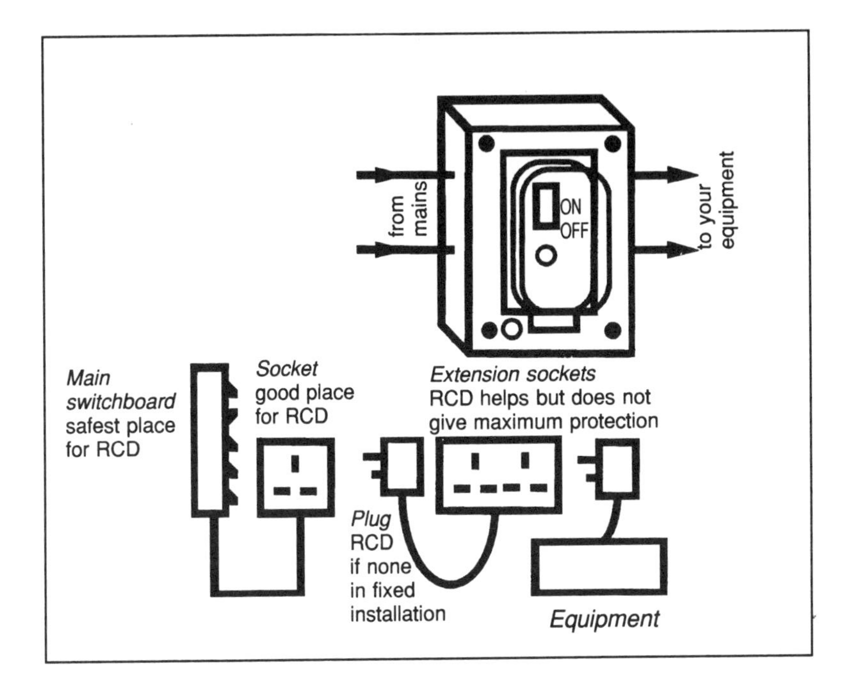

Fig. 11. Positioning an RCD.

chest muscles and upset your heartbeat rhythm. The higher the current, the more dangerous and quicker are the effects.'

It is advisable to carry an **RCD (residual current device)**, or **earth leakage circuit breaker**, which when connected to the electricity supply can detect even a very small leakage of current to earth. It will automatically switch off the supply if that leakage reaches 30 mA, the maximum you can normally take without any lasting effects.

RCDs have become essential to gardeners and DIY enthusiasts, and should similarly be so for entertainers who regularly ply their trade in venues where the safety of the electrical supply is not known. When you arrive at a venue it is not a bad idea to make your first job testing each socket on stage with a **Martindale socket tester**. This inexpensive device resembles a normal 13 amp plug with a set of three lamps on it showing the state of the socket.

If your RCD should switch off your power supply, it is a sign there is a fault which needs to be attended to by a qualified electrician. Never be tempted to by-pass the RCD for an easy life.

Earth loops

Any item which is mains powered should be either **double insulated** or correctly fitted with a **protective earth**. When a number of items are connected together you sometimes find that the screens, together with protective earths, form loops which create that annoying hum from your speakers. Do not be tempted to disconnect the earths to cure it. It can be eradicated by rearranging your cables so that power leads do not criss-cross with audio leads.

Using extension leads

This is another area fraught with danger for the unwary. Many view an **extension lead** as a simple thing to put together, consisting of a length of cable with a 13 amp plug on one end, and one or more sockets on the other. If only it were that simple.

If you really must make your own, make sure you are not using old cable which could have breaks in it. Check that it is the right size to power whatever it is you intend using it for. All electrical cable is colour coded:

- blue for **neutral**
- brown for **live**
- green and yellow for **earth**.

If your cable does not conform to this standard, throw it away. Remember, an extension lead should always be fully unwound when in use to avoid overheating.

Fuses

This is another area which many are guilty of infringing. When you purchase a 13 amp plug it contains a 13 amp fuse. The fuse is a safety device, and if the appliance develops a fault the fuse should protect both it and you. If the fuse is too highly rated it could damage the appliance, and perhaps you as well. When you fit a plug, make sure it contains the correct fuse. This is another requirement of the Portable Appliance Test which is discussed in more detail in chapter 11.

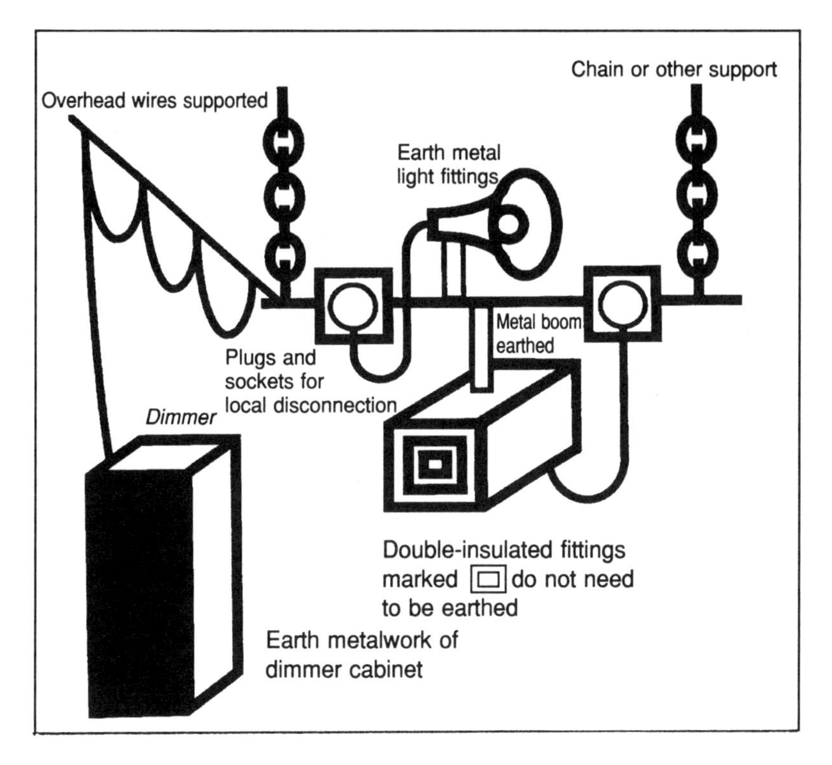

Fig. 12. Gantry lighting earthing points.

Lighting

Many mobile discos use teebars, goal posts and an assortment of trussing on which to display their lighting. When fitted to such rigs, each unit should be individually connected by plugs and sockets. **IEC plugs** and sockets, similar to those used on electric kettles, are ideal, and make it easier to isolate any single unit before changing the bulb or fuse. The metalwork of both the individual units, as well as the bar to which they are fitted, should be adequately connected to a protective earth conductor.

Generators

Outdoor gigs present their own set of problems for the unwary DJ. Damp grass, generator power, the over-use of extension leads from the domestic power supply, all make for a potentially lethal cocktail. Generators can kill just as effectively as mains power.

The main precaution to take is the provision of an **earth spike**, a copper rod driven some twelve to twenty-four inches into the ground. If the ground is very dry, a quantity of water should be poured over it to reduce the earth impedance to a low level. Keep all electrical connections off the bare grass.

SUMMING UP

- Your disco equipment represents a sizeable financial investment, so look after it carefully and it will serve you well for many years. Don't overlook periodical maintenance.

- When in transit, make sure equipment is packed carefully.

- Keep your toolbox fully stocked, and when you use the last fuse, piece of tape or whatever, replace it immediately. Add to it whenever you think of something else which might come in useful. You can never carry too many spares or tools.

- Most of your equipment is powered by electricity, so if you must fiddle about with the workings of an item, treat it with respect. Electricity can kill, so you must be safety conscious at all times. You have only one life.

CASE STUDIES

Lyn and the all-girl disco get technical

Lyn and her partners are aware that should a problem occur with the equipment at one of their gigs, they will get more abuse than their male counterparts because they are girls. Part of their reason for starting the disco was to show they are as competent as any male. When they purchased their equipment they made sure they got plenty of technical advice as to where fuses are located and their type, and they bought spares for their toolbox. Friendly DJs advised them to carry spare items of equipment, which they have done almost to the point of having enough for a second show.

Gary needs to be extra vigilant

Working with equipment which is supplied by the hire company for whom he works, Gary needs to be extra careful as he does not always get the same items. DJ friends have advised him to take a toolbox along to each gig, as they do, just in case. At the start of each gig he goes through a rigorous check of the equipment, noting where things are connected and where fuses are located. He realises he must be prepared for all eventualities, but has difficulty in persuading the hire company to provide a spare amplifier and so forth. Their attitude is 'The customer has not asked for, nor paid for it'.

DISCUSSION POINTS

1. What should a fully stocked toolbox contain?

2. Do you have a timetable or schedule of when your equipment should be checked?

3. Are your individual items of equipment sufficiently protected for transit?

4. Do you fully understand the importance of electrical safety, and do you know what to do if someone should receive an electric shock?

8
Working in Pubs and Clubs

Most of what you have learned so far has been directed at mobile discos. This is without doubt the hardest area in which to succeed as a DJ. The more glamorous market is clubs and pubs, an area associated with superstar DJs, more well-known and respected than their mobile counterparts.

ADVANTAGES AND DISADVANTAGES

The biggest plus point of clubs and pubs is that most of them have the necessary equipment already installed, so all you have to take along are your records and/or CDs. The work is regular and the fee will have been agreed before you signed the contract.

Club-goers demand a completely different music policy to that of the average mobile audience. They will appreciate the latest releases and like their music to be way ahead of the national charts.

The only disadvantage is the late hours which you will be expected to work; also your fee might not be as high as you would make with a mobile disco. You must remember, though, that you will not have any major expenses as you would with a mobile show, except for your records/CDs. Many club DJs manage to make a full-time living from it. Mobile gigs are confined to a Saturday night, sometimes Fridays as well, but clubs and pubs operate throughout the week.

SECURING YOUR FIRST BOOKING

Before you start touting for work it would be useful to notch up a little experience, and the easiest way to do this is as a roadie for a mobile. From this the club manager can see you have some sort of dedication and ability to do the job. The main priority of being a club DJ is to pull more customers into the establishment, to spend at the bar. It is not about airing your personal ego. The manager will be seeking the best available for his customers, so you will need to sell yourself.

Visiting the venue

When you visit a club or pub for this purpose, time your call for a quiet night early in the week. The manager will not have time to talk to you on a busy night, and if he is suitably initially impressed he will ask you to do a spot. If you make a good job of entertaining a few people in an almost empty room it will put you in a good position for a prime night later on. If you are offered a night, it will probably be on a dead night early in the week.

Negotiating your terms

If it is suggested you start off by doing such a night for free, think twice. The manager's reason for this may seem genuine, but after a month or two he might terminate your services and take on someone else for free. In this way, he gets all his entertainment free.

Politely explain you expect a fee of some sort, even if it is slightly less than that of the prime-night DJ. What you are seeking to achieve by being taken on is to turn that dead night into a popular one and boost attendances. Explain how you intend doing this by skilful use of press publicity, posters and word of mouth. It will cost you to supply such material, but it should be regarded as your initial investment in your future at that club or pub.

MANAGING AUDITIONS

There are two ways in which you might find yourself at a club or pub facing an audition:

- The cold sell when you turn up of your own choosing.

- Being sent, with others, by an agency.

An audition seeks to highlight how much you know about the job and whether your personality is suitable for the venue. It will also reveal what sort of music you will be playing. Auditions are not easy. It is difficult to work with unfamiliar equipment, and with no audience, but if you can handle this the manager will know you can handle anything.

Before, or sometimes after, this demonstration of your DJ skills, the manager will want to know a little more about you. Put together a portfolio starting with your CV of DJ-orientated skills, photos of yourself, showcard, and any other relevant information which tells the manager '*I'm the person for the job*'.

Doing your homework

Before going to any audition you must ascertain what sort of DJ the club or pub is seeking:

- Find out what sort of people go there, their age range and the sort of music they like.

- Arrive in plenty of time. Lateness, for whatever reason, may indicate you are unreliable.

- If selected you will become an important part of the club's team, so dress smartly. Take your records/CDs in a proper carrying case or brief case, *not* a carrier bag.

- You will probably be totally unfamiliar with the layout of the consol. It will be carefully explained and you will be given time for familiarisation. This is your big chance, so make sure you are totally happy with it before you begin.

- In selecting your half-a-dozen or so records/CDs, choose them carefully and rehearse linking them. Practise your vocal links and keep them interesting but tight.

- If you do make any mistakes, as you probably will, *do not panic*. Consideration will be given to the fact that you are using unfamiliar equipment; they will want to see how you handle it.

People go to clubs and pubs to relax and unwind. They expect to be entertained. Classing yourself as an entertainer, rather than just a DJ, raises your status to a level where if you wished you could apply for membership of Equity.

DEVELOPING YOUR STANDING

Once appointed as a DJ in a club or pub, you become a responsible member of the team. Each member of that team has an important role to play in the continued success of the place. Club-goers are a fickle lot. If anything spoils their enjoyment they move away to another establishment in droves. Part of your job is to keep them with you by providing entertainment which is always new, fresh and exciting. If you fail in this, you will be replaced very quickly. It is true to say 'You are only as good as you were yesterday.' How can you build up your standing?

Promotional visits

As a club DJ, you will be eligible to receive copies of promotional records, as discussed in chapter 4. Part of the promotion includes the artiste visiting as many clubs as possible to perform his/her new release, usually free of charge. If a particular new release is going down well, seek the manager's permission to invite that artiste along and arrange for local press coverage of the event.

Competitions and promotions

Through the trade magazines you will learn of nationwide competitions and promotions, all seeking venues in which to stage their event. These will boost the numbers at your club or pub if used. They usually cost, so you have to decide whether it's worth it.

Inventing theme nights

Less expensive are theme nights. Your diary is full of 'special days' so devise a 'special night' to suit. The more obvious are St Valentine's, Halloween and Guy Fawkes Night, so get your thinking cap on and come up with ideas such as a St Trinians Night at the end or start of a school term.

On other nights bring in little competitions, such as one I once did by inviting club-goers to bring something which suggested a record title. Amongst the entries I got was a road lamp on a roller skate (Travelling Light—Cliff Richard), and a toilet pan complete with seat and cover. When you raised the cover, it revealed a staircase. What was it? Of course, Stairway to Heaven by Led Zeppelin!

HANDLING SPECIAL OCCASIONS

As the club's DJ you are an ambassador for them, due to the high profile of your job. When you invite someone special along, treat them as courteously as you would at home, no matter whether it's a recording artiste, the organiser of a competition who has called in to check you out, a local celebrity or even a potential new DJ.

- Warn the door staff to look out for them and escort them to you when they arrive.

- Advise the door staff to show your guest(s) where to park their vehicle safely if there is not an official car park.

- Introduce your guest(s) to the manager, then take them to the bar and introduce them to the head barman. Let him know whether it is in order to offer free drinks.

- If the club has someone who works as a host, politely hand your guest(s) over to him/her. By now you will be required on stage.

INCREASING YOUR ROLE

Eventually you could rise to the dizzy heights of being the prime night DJ, working either Friday or Saturday nights, or both. So far, you have only been responsible for your own particular night. You are now the senior DJ and, as such, those newcomers struggling to make their mark on the dead nights earlier in the week will be looking to you for inspiration.

This is when you could start to assume the role of **promotions co-ordinator**. By now you will have earned a certain standing and amassed numerous contacts. Use everything at your disposal for the further good of the venue. If, for example, your town has an annual carnival, why not get the club or pub involved in it for publicity? Your skills must expand from those of an ordinary DJ to those of a public relations consultant.

Making sure of your position

The reasoning behind expanding is to secure and stabilise your position. As news of your achievements start to spread you may find yourself being head-hunted by other clubs. This of course is very flattering, and it is entirely up to you whether you accept any such offers or remain loyal to your existing employer. You may be offered more money or better conditions, but it really is a matter between you and your conscience. Whatever you decide, talk it over with your manager. If he values you highly enough he may be able to offer you more money.

ADOPTING A STAGE NAME

One feature of being a club DJ seems to be that many adopt strange-sounding names. You might wonder if this is necessary, indeed essential, for your success. Not at all. There is absolutely no evidence to suggest that by changing your name you will have any greater degree of success.

Likewise, there is no reason why you should keep the name you were christened with, especially if it doesn't roll off the tongue easily. I am not over-enamoured of unimaginative names like DJ Mix, Scratch Doctor, Professor Beat and so forth. If you achieve any success in your career you will be stuck with your adopted name for ever, so choose something which will grow with you. You may move on to other media as you progress, and though DJ Mix suited a 16-year old mixing DJ, how does it sound for a 25/30-year old working in radio or writing?

If it is your ambition to make a life's work of being a DJ, and you intend changing your name, then before settling on your new one make sure it does not already exist by checking it out in the *Blue Book of British Broadcasting*, published by:

Telex Monitors Ltd
210 Old Street
London EC1V 9UN
Tel. (0171) 490 1447.

It is costly so it might be advisable to consult your local library.

Having acquired a new identity, it makes sense to protect it by Deed Poll. It is not essential, but if someone else should also adopt it they might be using it to your detriment.

To change your name by Deed Poll:

- Visit a solicitor; it does not need to be a specialist.

- Sign a sworn declaration, and pay a fee of about £10.
- Wait a couple of weeks, and you will be formally notified you have a new identity.
- Good-bye Vladimir Kzatchowski—hello Dave Sinclair!

CHECKLIST

- Having decided your niche as a DJ lies in the world of clubs and pubs, put together a portfolio with which to sell your services.
- Visit potential venues early in the week when they are quiet, and be prepared to spin a few records/CDs to show what you can do.
- Don't be tricked into working for nothing.
- Once accepted, consolidate your position by organising special events and competitions with a view to taking on all publicity functions.
- Decide at the outset whether you intend keeping your own name or adopting a stage name.

CASE STUDIES

Phil goes for club work

Phil, aged 25, has been running a very successful mobile disco for over ten years, but is now seeking a fresh direction. He decides to sell his equipment and move into the more glamorous and glitzy world of club work. He has put together a comprehensive portfolio, including photographs of himself dressed in different ways. The manager spots one of Phil in a dinner jacket and says that is the image for a new night he has been contemplating for some time. Unbeknown to Phil, this has been a stumbling block. The manager could not find the right DJ, who could dress up and look comfortable in a dinner jacket.

Les gets things going at the pub

Les, aged 19, is a regular at his local pub. On one particularly wet Tuesday evening he was commiserating with the landlord at the lack of customers. Les suggested that what was needed to draw people in was some sort of entertainment, such as a disco which would be far cheaper than a band. The landlord explained he could not afford such a thing as it might not work. Les suggested the landlord hire the equipment on a couple of evenings, and he would play the role of DJ for free to start with. If the idea worked, the pub could then purchase the necessary equipment and negotiate a fee for Les.

The landlord agreed, the idea took off and Les slowly developed it in several ways. The pub soon became the in-place in town.

DISCUSSION POINTS

1. How do you see the role of the DJ in clubs and pubs?

2. Compile a suitable portfolio with which to sell your services.

3. What scope is there for continued success in this sphere?

4. If you decide to change your name, compile a list of suitable ones, and check them out as widely as possible.

9
Working in Radio

Many DJs who start by working in disco soon aspire to working in local radio. It is assumed a DJ is a DJ, each playing music to a large audience. Whilst this may be true in part, the majority of radio **programme controllers** do not see disco DJs as potential radio presenters.

REQUIREMENTS AND QUALITIES

Radio is a medium which demands many varied skills.

- You must be prepared to work hard.
- You must be a fast thinker.
- You need to be resourceful.
- You will need a clear, pleasant speaking voice.
- You must be able to communicate with your listeners.

Listeners like to feel the presenter is talking to them on a one-to-one basis. You must, therefore, be natural, sincere and sufficiently experienced in life to relate to your audience. As part of the radio presenter's workload involves meeting listeners at roadshows and public appearances, some disco experience will be useful.

Where do I get training?

Before contemplating a move to radio, you will need some form of training to acquire the necessary skills. Such training cannot be found in your local nightclub or pub. It cannot even be found with a mobile disco, but it *can* be developed in each of these areas once you have a grasp of the basics. The only place to learn these skills is in a radio station itself. It would appear to be a Catch 22 situation—you cannot work in radio unless you have experience, and you cannot get that experience unless you work in radio!

There are several options open to you, most of which gladly accept unpaid volunteers. These include:

- hospital radio
- in-store radio
- campus radio
- and community radio.

None of these should be viewed as training schools as such. Each is a separate entity in its own right, but if you are accepted by any of them, in whatever capacity, you will learn much. When you do apply for work in a 'proper' radio station the experience you gained will put you in good stead.

HOSPITAL RADIO

This is a radio system which operates on a closed-circuit basis to one or more hospitals. It is relayed on BT landlines similar to your telephone line, but at a higher quality. Hospital radio is for the benefit of the patients, and has been shown to be of tremendous therapeutic value to morale.

Is it real radio?

It is real radio in the sense that most hospital radio studios are equipped almost to full IBA specification. They have a programme controller to oversee programming and overall output. Most volunteers strive to perfect their technique and professionalism. Many of today's top radio *and* TV presenters began in this way.

How do I start?

It's as simple as ringing your local hospital and asking whether they have such a station. If so, they'll gladly give you the contact name and telephone number.

As a volunteer you will be expected to do all manner of tasks, all essential to the successful running of the station. You might not immediately be asked to present a programme, but whatever your role you will learn a lot. Hospital radio stations operate as a team. Membership is a sign of commitment, not something you do as

Fig. 13. A hospital radio studio (Radio Swale).

and when you feel like it. It's a commitment to the patients and to the rest of the team.

What might I be expected to do?

- **Technicians** constantly repair, maintain, buy and even make items of equipment.

- **Record librarians** keep indexes up-to-date and maintain the filing system of records, CDs, tapes and so forth.

- **Ward reps** visit the patients to get record requests for the presenters.

- **Freelance presenters**: any member of the team can go out with a portable tape recorder to produce an item for possible broadcast. Such items might be a backstage chat with an artiste at the local theatre, the town's MP on a new political issue, or an item on an anniversary such as VE Day, Remembrance Day or Trafalgar Day. The scope is enormous, and it is an excellent way to start as a broadcaster.

Who pays for all this?

There is usually a small membership fee, but the hospital's League of Friends, that band of people dedicated to supplying those things to the hospital not covered by the NHS, generally help out. Most hospital radio stations organise a series of fund-raising events throughout the year.

What skills are necessary to present a programme?

Exactly the same as on any other radio station:

- an ability to communicate
- slick presentation
- working within precise time limits
- broad musical tastes.

Music policy

Much of hospital radio's music output is record requests, mostly middle of the road standards and favourites. Remember, most of your listeners are sick people, so steer clear of any songs which include lines about death and dying.

Other taboo topics

The good hospital radio presenter should beware of such clichés as time checks and weather forecasts, neither of which are of importance to the bedridden patients. The biggest gaff might be at the end of your programme to say, 'I look forward to the pleasure of your company next week . . .'. The patient will doubtlessly be hoping to be out long before then.

Contacts

There is a National Association of Hospital Broadcasting Organisations, known as NAHBO, which was inaugurated in 1970 as a national body to encourage and assist the formation of hospital broadcasting services. NAHBO can advise on registering as a charity, studio design, equipment choice and a model constitution. They can also provide technical and admin advice. Of the 200-odd hospital radio stations in the UK, 190 are members. They can be contacted at:

Milne House
1 Norfolk Square
London W2 1RU
Tel. (0171) 402 8815 or (01324) 613744.

CAMPUS RADIO

Should you be lucky enough to get a place at university or college, you will find that many have a campus radio station. This is another excellent stepping stone towards your career in broadcasting. Unlike hospital radio these stations operate via a low-powered transmitter, and the signal can be picked up on an ordinary radio within the confines of the campus. It is loosely described as radio for students, run by students.

Contacts

Further information on campus radio can be obtained from:

The Student Radio Association.
Tel. (01602) 513617.

IN-STORE RADIO

A third way of gaining useful radio experience is through in-store radio. These operate on the same closed-circuit principal as hospital radio stations, but for the benefit of shoppers. The system can work as well in a single departmental store as it can in a shopping precinct.

Your audience will be young, trendy, fashion-conscious people, so your music will need to appeal to them. A large part of the job will be to voice frequent live commercials advertising the store's products, so you will need a degree of creativity to put these together.

Several large stores already have an in-store radio station, notably Top Shop, HMV Records and Virgin Megastore in Oxford Street, London, and ASDA supermarkets.

How do I break into this market?

You will probably have about as much chance of getting into one of these stations as you would into a local BBC or ILR station. Your best move might be to approach a large store or shopping arcade in your own town, and ask to set up a new station. Be prepared to sell the idea and fully explain the many advantages of such a radio station.

COMPILING A DEMO TAPE

When you feel you have sufficient experience with which to make a serious application to a BBC, ILR or **community radio** station, it is time to consider putting together your application package. At the heart of this will be a **demo tape** demonstrating your skills. (If you are wondering what community radio is, read on, all will be explained.)

Applying for work on any radio station is a critical operation, as your application has to immediately catch the eye of the programme controller. Your tape will be one of many he receives each day. No matter how good you are, if your application does not *look* good it will not even get to the stage of being looked at, let alone listened to. Remember that the programme controller also receives many applications from seasoned veterans.

Initial considerations

Radio presentation is a highly professional art which involves more than just playing music. You'll need wide and varied talents, such as being able to interview all manner of people from all walks of life. You'll need to be able to put together reports, edit tapes and read the news. Any music you do play will need to be compiled in a well-balanced way to appeal to the majority of your listeners. The order of the day is *flair* and *originality*. Your demo tape needs to show:

- your vocal quality
- personality
- reading ability
- maturity
- general attitude.

Format

There are several formats you could choose on which to record your demo:

- open spool tape
- cassette
- DAT
- minidisc.

The most popular with programme controllers is the ordinary domestic cassette. Avoid using a standard C60/C90, even if it is a good quality brand. Buy a professional short length, or computer tape of about ten minutes duration each side.

Your demo should be a maximum of five or six minutes in length, no longer. The programme controller has many such tapes to listen to, so it's important to seize his attention within the first few seconds, otherwise your tape will be consigned to the reject box. The start of your tape is the most important part, and the most difficult to get right. It is imperative it both looks good and sounds good.

Content

Start off by stating your name, address and telephone number. A good way to begin is by playing an instrumental piece of music, over which you can read a weather forecast, traffic report, or rundown of what's to come.

You then need to show as many of your talents as you can in as short a time as possible. Anyone can introduce music, so include as much speech as possible by way of a quiz, a 'what's on' diary, or a news bulletin. You can, of course, introduce a couple of records/CDs to show your musical tastes and how you can introduce them in an original way. All music will need to be **topped and tailed**.

Topping and tailing
After the first three or four seconds of music the record/CD should be **faded out**, or **cut**, using the **pause button**, then faded back in or the pause button lifted, three or four seconds before the end. This is not quite as easy to do as it sounds, if you want a smooth edit. You should be aiming to make your edit firstly on an **outro beat**, followed by one on an **intro beat**. Practise this carefully to get a smooth edit, otherwise it tends to jar the listener, whom we don't want to upset.

News bulletins

You will need to show there is a serious side to you, and this can be achieved by reading a **news bulletin**. For the purpose of your tape, take some clippings from your local newspaper and re-write the stories in a short, sharp, but informative way. The news bulletin is very important and should be read straight and factually. Never inject your own thoughts or comments on any of the stories. Do not race through it making it sound like something you want to get out of the way. Keep it flowing in a normal, conversational tone, not with a stilted approach.

Before recording this passage read it through carefully several times to familiarise yourself with it. If certain sentences seem a little long, split them into shorter ones. Using commas as breathing marks helps the flow.

Part of your tape might include an interview, something mentioned in chapter 6. In a radio environment this can include phone-ins. As the presenter, be in control at all times. You instigated the interview; you can terminate it.

Ending your tape

To conclude your tape play a piece of instrumental music, perhaps the same which you started with for continuity, and over it thank the programme controller for listening, perhaps adding how much you look forward to hearing from him/her.

Now listen to your tape critically several times, asking yourself if it really is the best you can manage. Record two or three different versions and make your final choice from these.

Next, listen to your chosen station to see how your tape fits in with their programming. It would be pointless, for example, to send a Radio 1-styled tape to Radio 2.

The complete package

Add to the tape your CV, and a recent head-and-shoulders photograph of yourself. Type out a covering letter, then sit back and wait patiently.

WORKING IN COMMUNITY RADIO

Getting into radio is a difficult business. It is highly competitive, and many people suggest it is a case more of *who* you know than *what* you know. An easier option would be to get involved with one of the new community radio stations which are springing up all over the country.

What is community radio?

These stations are truly local radio as it should be, and lie somewhere between hospital radio/campus radio/in-store radio and the local BBC and ILR stations. Sometimes referred to as 'special events' radio or 'festival' radio, they are fully legal, licensed by the Radio Authority, and can broadcast for up to twenty-eight days at a time. They have to meet professional requirements of broadcasting. Their coverage area is typically a radius of not more than 25 miles.

These stations operate on an experimental basis, with a view to creating a good impression and making a significant impact both on the community they serve and on the Radio Authority itself. Their ultimate aim is to make a good case for being awarded a permanent licence to broadcast as a community station.

By getting involved not only do you get valuable on-air experience broadcasting and using professional equipment, but you also play an important part in helping the station secure a permanent licence later on. As a volunter in the experimental twenty-eight day stage it is unlikely you would be paid any sort of fee, but if the station were to gain a permanent licence it would not only secure you a slot, but probably a fee as well.

Comparisons with ILR stations

The major difference between an ILR (commercial radio) station

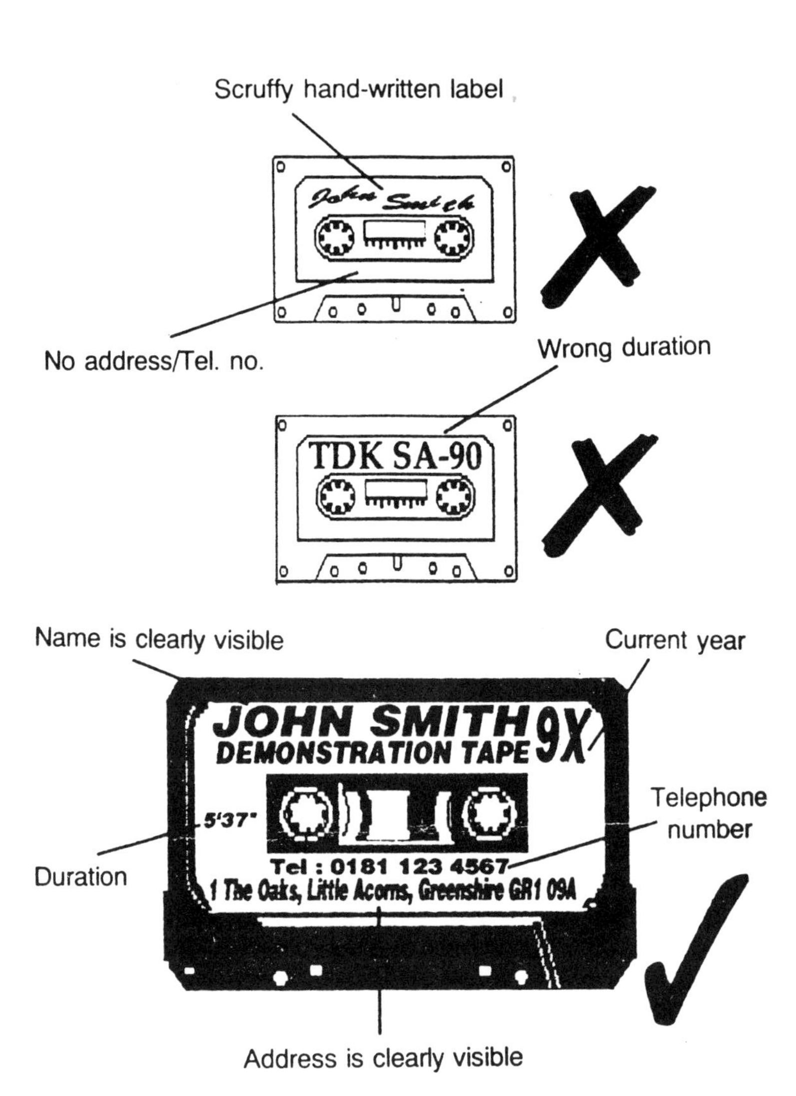

Fig. 14. Demo cassette label.

and a community station is the size of their respective audiences. Both are on-going commercial concerns, depending on advertising revenue for their survival. Each broadcasts a general mix of music and local news and information, regulated by the Radio Authority.

Your first major break?

If you have had some experience as a DJ or presenter in hospital radio, campus radio or in-store radio, you stand a very good chance of being accepted. It helps if you live in the town the station is sited in as you will have an appreciation of the area, its problems and its people.

Contacts

The Community Radio Association
The Media Centre
15 Paternoster Row
Sheffield S1 2BX
Tel. (0114) 279 5219.

Community Radio Association (London Development Unit)
Lambeth College
Belmore Street
London SW8 2JY
Tel. (0171) 738 8788.

BBC AND ILR RADIO

Having gained suitable experience and put all your skills to good use on a community radio station, your ambitions may next drive you to seek an opening on a local BBC or ILR station. Do not be tempted to try going straight to the top by applying to national BBC. Their staff usually come from one of the many local BBC or ILR stations. This is where you must begin, and it would be well to study Bernie Simmons' excellent book *How to Get into Radio* (How To Books) first. It is a complete subject in itself.

CHECKLIST

- If it is your burning ambition to become an all-round radio presenter, you will need to gain experience both in broadcasting and in using professional equipment first. This can be found in hospital radio, campus radio and in-store radio.

- Once you feel competent to present a programme professionally, apply for a slot on a community radio station.

- This background should then put you in good stead for an application to a local BBC or ILR station.

CASE STUDIES

Jon starts with hospital radio

Jon, aged 23, was unable to put into words exactly why he wished to become a broadcaster, it was just something he wanted to do. He arrived at the hospital radio studio one evening to look around, was fascinated by all he saw and was told, and asked to join straight away. The team were instantly impressed by Jon's personality and enthusiasm, and one presenter even offered to take Jon under his wing to show him the ropes. In time Jon's infectious personality caught the imagination of the patients and he proved to be a popular broadcaster.

A year after joining, community radio came to town and several of the hospital radio presenters, including Jon, were accepted. This proved to be his springboard to bigger and better things.

Simon proposes an in-store radio station

Simon, aged 35, had some ten years' experience in hospital radio where he had got a sound grounding in a DJ's assorted skills. He left to set up a disco roadshow, which flourished.

When a new shopping arcade was built in the town centre he could foresee it playing 'muzak', that awful, bland background music system. Simon arranged a meeting with the developers of the arcade and found his fears on the background music system were well-founded. He had prepared a presentation of facts, figures and a tape showing the many advantages of an in-store type of radio station. The developers were suitably impressed and promised to seriously consider the proposal.

DISCUSSION POINTS

1. Map out your plan of attack, showing what qualities you need to be considered for a job in radio.

2. Draw up a schedule of items to be included on your demo tape.

3. List the differences between a local BBC and ILR station.

4. What role does a community radio station play within the community?

5. List the differences between a hospital radio, campus radio and an in-store radio. Why does each exist and what is its role?

10
Mixing in Your Bedroom

Not every potential DJ has aspirations of playing to an audience. Some are quite content to buy the necessary equipment, set it up in their bedroom and put together tapes for their own personal pleasure. Nothing wrong with that. Many of today's top **mixers, remixers** and **producers** started in this way. If you do too, and become proficient and want to forge ahead, there are many avenues open to you.

LOOKING AT THE BASIC EQUIPMENT

Start-up packages can be purchased from most retailers for as little as £250/£300. For this you could expect to get a pair of direct drive turntables, a basic mixer, a pair of headphones, all necessary connecting leads and a pair of slipmats. Naturally the more you have to spend the better the quality of equipment you can expect to get.

As a mixing DJ you will not need a microphone, but you will need to connect your mixer to either a cassette recorder or reel-to-reel recorder. A pair of **monitor speakers** on which to listen to your final mixes might also be useful. If used, remember you'll need to firstly connect your mixer to an amplifier, unless of course you purchase those monitor speakers which have integral amplifiers.

Editing

The most simple and basic way to mix is by engaging the pause button on your tape recorder. This can be a very hit-and-miss affair until you perfect the technique. It is far from the ideal way in which to produce a mix, but it's a start which, if nothing else, will give you experience in arranging and structuring a mix.

Another way is to use the **cross fade** facility on your mixer if it has one, which most simple mixers designed for bedroom mixing do. Under normal operating conditions the fader is kept in the centre of its track. When you wish to fade out, say the left channel and bring in the right one, you push the fader to the far right. When you wish to bring back the left channel, return the fader to the centre or far left.

Pause editing
A point worth remembering when pause editing is that your tape recorder might be a model which has **logic control buttons**, or **touch pads**. These do not react as quickly as the older models which had **tab switches** or **piano key controls**.

DEVELOPING YOUR ROLE

Once you have successfully mastered the art of structuring a mix you will no doubt think that a certain piece of music you have used would have fitted better if it had been arranged differently. This is the stage where you move from being a mixer to a remixer.

Much of the music remixed by today's megamixers is based on the **house music** style of the 80s. It is a style of music evolved from computers. Where at one time music was written on musical staves and handed to the individual musicians, it is now converted into numbers and fed into a computer.

SEQUENCING WITH A MIDI

In upgrading your home studio you will need a suitable computer along the lines of an Atari, which has **MIDI** sockets (**musical instrument digital interface**), a system which allows you to connect electronic musical instruments such as **sequencers, samplers, drum machines** and so forth to your system.

With your computer in place you will first of all need some **music software** to create your sequencer. Industry standard sequencers include C-Lab Notator and Steinberg Cubase. These can be an expensive item to purchase, so consider buying second-hand from magazines on computing and recording, which your local newsagent or library might have.

There are various electronic musical instruments you could connect to your MIDI.

Sampler

A sampler is the most powerful music manipulation device you can get your hands on. It is capable of replacing all other instruments, and can be best described as a computer which records and stores sounds in the form of numbers.

There are numerous models available at varying prices, but the more expensive the item the more facilities it will offer. Again, look out for secondhand bargains. Brand names to look out for include Akai, Mirage, Casio and Yamaha.

Drum machine

This is an expense you might be able to forego as your sampler could quite easily take on this role. You may find, however, that you are using all the **memory** or **outputs** of your sampler, so a separate drum machine could be the cheapest way of splitting the workload.

Synthesiser

Adding a few melodic breaks to your mix is another option you may like to consider, even if you cannot play the keyboard with any conviction. A crafty way around the problem would be to use just the memory of the synthesiser, known in the trade as an **expander**. Your local high street music shops will be able to offer a wide selection of suitable keyboards, but make sure they can be connected to a MIDI. Not all of them can.

THE STUDIO

Up to this point your studio might have the following equipment:

- sound mixer, preferably a six-channel model but a four will do
- a pair of direct drive turntables
- computer
- musical instrument digital interface
- sampler, drum machine and synthesiser
- tape recorder, preferably reel-to-reel, of professional quality.

FURTHER EXPANSION

If earlier on you purchased a very basic sound mixer, and you are now seeking something with more inputs, it is a sign that you could be seeking a **mixing desk**, rather than a basic mixer. Such a desk will give you far more control over each channel by way of **volume** and **equalisation** (or tone control), and the possibility of adding **special effects** such as digital delay, reverberation, sound gates and so forth to any selected channel.

And there's more

Your bedroom studio is by now capable of producing some good quality and very useable material. You could next progress to **multi-tracking recording** which stores four, six, eight, twelve or sixteen different tracks on to the same tape, but this is starting to go into the realms of the professional recording industry.

Contacts

This is a long way in a short time from the initial, basic start-up package. Home recording is a pastime which knows no bounds, and if you want to get thoroughly immersed in the subject the School of Audio Engineering runs several courses, both full-time and part-time, on various aspects:

United House
North Road
London N7 9DP
Tel. (0171) 609 2653.

A useful textbook to read is *How to Set up a Home Recording Studio*, published by:

PC Publishing
4 Brook Street
Tonbridge
Kent TNP 2PJ
Tel. (01732) 770893.

Going further

- As with a mobile disco, the bedroom studio can be upgraded and added to as and when finances allow. Virtually anything which makes a noise can be used in your mix, even simple things such as video games, musical toys and books, dictaphones.

- You can spice up your mix by using additional samples and sound effects. Make sure they become an integrated part of the mix, not something laid on top for the sake of it.

- Editing your mix is another important point to bear in mind. You don't necessarily have to wait until a break comes along before going into your next track. It needlessly pads out the mix, making it too long. Megamixing is not about just the links, but the bits in between as well.

- When you become a megamixer, you start to become a producer and, as such, you start to listen to other peoples' work more critically. This is what will inspire you for future mixes. Most mixers agree the hardest part of any mix is the beginning. Adopt the saying 'Start as you mean to go on'; if the beginning is weak and has a lack of direction, it will be very difficult to progress with any sense of conviction.

USING THE FINISHED PRODUCT

Having spent so much time, money, trouble and effort to achieve the standard of mixes you have, it makes sense to use them in some way. You could advertise them for sale, but you will soon encounter problems over copyright. A better option might be to offer them to one of the companies which specialise in producing such mixes and remixes for use by subscribing DJs. The two major distributors in the UK are Music Factory and the Discomix Club. Before you send a tape in:

- Telephone first to check the concept of your mix.

- Write down a full track listing with the running time of the mix plus your name, address, and telephone number.

- Your mix should be recorded on either a DAT tape or 1/4in reel at a speed of 15 ips. Cassette tapes are fine for demonstration purposes only.

CHECKLIST

- Begin your studio with either a start-up package or separates consisting of a mixer, two turntables, headphones and a tape recorder.

- To listen to your finished product add an amplifier and a pair of speakers.
- As you become more ambitious and creative, add a computer which will take a MIDI.
- Connect to your MIDI any electronic musical instruments you wish.

DISCUSSION POINTS

1. How does the role of the bedroom mixer differ from that of any other sort of DJ?
2. What outlets are available for this work, and can you think of any more?
3. How does a mixer differ from a producer?
4. What are the advantages of a mixing desk over a basic mixer?

Throughout this chapter it has been assumed you will be mixing using records. It is possible to mix using CDs, but they do have a slightly slower take-up time. Models to look out for are the Pioneer CDJ500, Denon 2000F and Vestax CD11. Each of these incorporate vari-speed, pitch control and a sampling facility. Two others which are slightly less expensive, but still worthy of consideration, are the Neumark CD7020 and Gemini. Costs vary between £600 and £1,000, so shop around.

11
Meeting Legal Requirements

As a DJ, no matter what your status or commitment, you have certain legal obligations. Non-compliance could lead to a hefty fine and/or imprisonment. It clearly states on each and every record, CD and tape that you should not play it in public, unless licensed to so do.

ACQUIRING LICENCES

Licensing is another of those grey areas which can cause much concern to newcomer DJs. Licensing exists to protect the **copyright** of writers, publishers, performers and record companies. There are four organisations which may concern you:

- Performing Rights Society, or PRS
- Phonographic Performance Ltd, or PPL
- Video Performance Ltd, or VPL
- Mechanical Copyright Protection Society, or MCPS.

It is the job of these organisations to collect any fees due.

Having purchased a record, CD, tape or whatever, you can only play it in the privacy of your own home. You could almost be excused for thinking that once you have made your purchase the item is yours to do with as you see fit. In reality you never actually purchase the complete item. Various copyrights exist, protecting everyone's interests. This is a situation which does not exist in any other area of purchasing, except perhaps in publishing. Take a brief look at each organisation, and ascertain what applies to you and where to go for more information.

Performing Rights Society

This is a non-profit making association of composers, authors and music publishers which exists to collect and distribute royalties for the public performance and broadcasting of its members' music. PRS charges a fee to every venue where copyright music is played. This includes:

- clubs
- pubs
- hotels
- church halls
- cinemas
- factories
- and even shops where a radio is played for the benefit of customers and/or staff.

The PRS normally licenses the venue, but where there is no annual licence the promoter of each event is held responsible. A blanket licence is available for a nominal sum to cover venues not normally used for the purpose of music and dancing, such as private houses, barns, marquees and so forth. This licence does not cover venues which should already have a licence issued by the Performing Rights Society.

The bottom line for all DJs is to make sure all bookings are covered by a legally binding contract which includes, amongst the small print, the phrase 'the person who is booking the discotheque is responsible for obtaining all necessary licences and permissions'. If you are a member of a recognised DJ association, and use one of their official contracts, it will probably be included. This subject is dealt with in more detail in chapter 12.

Contact

Performing Rights Society
29–33 Berners Street
London W1P 4AA
Tel. (0171) 580 5544.

Phonographic Performance Limited

PPL protects the interests of artistes and record companies. You are prohibited from playing recorded music in public, but the Copyright Act 1956 does not define exactly what a 'public performance' is. It can be best described as a public place, even if it is for members only, irrespective of whether an admission charge is made or not. For example, a wedding reception or similar family function is deemed to be a private function, even if held in a public place such as a village hall or hotel.

PPL collects its fees from the owner of the venue. Mobile disco DJs who organise their own function can obtain a one-off licence for a single event at a modest fee. A licence is still necessary even if you only play promotional material.

Contact

Phonographic Performance Ltd and Video Performance Ltd
Ganton House
14–22 Ganton Street
London W1V 1LB
Tel. (0171) 437 0311.

Video Performance Ltd

Like PPL, VPL is a non-profit making organisation established by British producers to issue licences for the public use of videos featuring their work. Venues which show moving pictures for profit must also be licensed as a cinema by the local authority in accordance with the Cinematograph Amendment Act 1982.

Address as for PPL above.

Mechanical Copyright Protection Society

The **mechanical copyright** is the right to record music onto whatever medium you choose. To enable the copyright holders to receive fair remuneration for their work, they receive payment from the record companies in return for permission to record their work. Payment is also due from any individual who wants to re-record copyright music.

Contact

Mechanical Copyright Protection Society
Elgar House
41 Streatham High Road
London SW16 1ER
Tel. (0181) 769 4400.

The role of the British Phonographic Industry

There is one other body which helps the preceding four organisations, but with which you will not need to get involved: the BPI

MOBILE DISCOTHEQUE OPERATIONS AND THE PHONOGRAPHIC PERFORMANCE LICENCE

Phonographic Performance Limited, Ganton House, 14—22 Ganton Street, London W1V 1LB Telephone 01 437 0311 Telex 268610 PPLVPL

INFORMATION FOR MOBILE DJ's

Introduction

The Copyright, Designs and Patents Act 1988 provides a copyright in the public performance of sound recordings (records, tapes, compact discs etc). A copyright cannot be seen or touched but is, effectively, a piece of property. It is assigned to Phonographic Performance Limited (PPL) by the majority of record producers so that it may be used practically and efficiently. This helps users obtain access to a very large repertoire of recordings. PPL is a non profit making Company and all of our income (less running costs) is given to the record producers and artistes.

When sound recordings are publicly performed our authorisation is required. This is given, where possible, in the form of a licence.

To help you decide whether or not our licence will be required we have set out the most common questions with their answers below. If something is not covered please contact us.

When do I need a licence and what does it cover?

In general, whenever sound recordings subject to our control are played in public in the UK. A licence should be obtained whether you are professional or amateur, large or small. It covers the use of the great majority of records, tapes, CDs etc. available and it is very unlikely that any discotheque could be held without using PPL controlled recordings.

Can I get an annual licence?

Yes, PPL can issue an annual licence to individual mobile DJ's. This will ensure that you are legally covered for the public use of sound recordings at any number of single events held during a twelve month period.

Does this cover me for everything?

No. If you have a residency in a pub or club or you are playing at consecutive events held in the same venue, your licence will not cover you. Details of residencies and regular bookings should be passed to PPL and we will arrange licensing with the promoter of the events or the owner of the premises. You will also not be covered to play in discotheques, nightclubs or similar venues as the owners of such venues need to apply separately to PPL for our licence. To ensure you are not infringing PPL's copyright in these venues, please pass any information to us so that we may advise you further.

Do I need a licence for weddings, birthdays or other family anniversaries?

No. PPL's licence is not required for family or domestic events, even if the event is held in a public hall.

Do I need a licence for a private event held by a sports or social club or a firm?

Yes. Even though admission is restricted to club members or staff, such events are public for copyright purposes and a licence must be obtained. However, if regular events are held by a sports or social club or firm please provide details to PPL and we will arrange licensing with the club concerned.

Fig. 15. Questions and answers about licensing

I have been told that the hall is already licensed. Do I still need a PPL licence?

PPL does not issue licences to halls, only to people responsible for the public use of sound recordings. Other licences may be required from the Council and the local Justices. These are not concerned with copyright. The hall may have a licence from the Performing Right Society (PRS). This is concerned with the rights owned by composers and music publishers and is quite separate and additional to that required from PPL.

I am promoting my own events. Does the mobile DJ licence cover me?

No. If you are promoting your own events you will need to secure a different licence based upon the number of events you are likely to hold during a year. The mobile DJ licence only covers you for single events where you are employed by a third party. Please contact PPL for an application form if this is required.

What happens if I do not get a licence?

When infringements occur (that is, when our members' recordings are used without authorisation), our normal policy is to secure an injunction in the High Court. Such an injunction prevents any further use of sound recordings subject to our control. It has no time limit. Legal costs and damages may be payable.

Can I obtain a licence from you to re-record from disc onto tape?

No. Permission from the individual copyright owner is required. We shall supply further details if required.

What if I use imported recordings?

Most foreign recordings are copyright protected in the UK. International licensing agreements often exist between our member companies and the foreign producers; such recordings may be considered subject to our control.

Can I get free sample records from you?

No. PPL is not a manufacturing company. Please note that if you are lucky enough to receive free promotional copies from our member companies the copyright restrictions still exist. Promotional copies are issued on the understanding that their use will be subject to the usual licensing arrangements.

Can I register the name of my mobile with you?

No. PPL is only concerned with the copyright in sound recordings. We therefore have no such register of names and are not in a position to offer protection against other mobiles using the same name. You may, however, be able to register your business name with the appropriate authorities and we suggest that you contact a solicitor for advice.

If you wish to secure PPL's Mobile DJ licence please complete the enclosed application form and return it as soon as possible to:
PHONOGRAPHIC PERFORMANCE LIMITED, GANTON HOUSE, 14–22 GANTON STREET, LONDON W1V 1LB TELEPHONE 01 437 0311 FAX 01 734 9797

or British Phonographic Industry. It is their job to stamp out illegal home taping and commercial piracy.

How long does copyright exist?

Copyright exists on every piece of music for the duration of the life of the composer and for fifty years after his or her death. There is a separate copyright for records, CDs, tapes and videos which lasts for fifty years from the date of their first publication. Performance of copyright music in public places is covered by the Copyright Act 1956.

PORTABLE APPLIANCE TESTING

Under the Health and Safety at Work Act 1974, it has become the responsibility of both employers and employees, including the self-employed, to ensure as far as is reasonably practical the health and safety of all persons in the workplace. Nowhere is exempt, and it applies equally to shops, offices, factories and places of entertainment. It also covers the home if that is where your employment is based.

Staying safe

The electrical safety of **portable appliances** such as disco equipment became law in April 1990. The regulations require that all appliances which have a plug on the end of an electrical lead must be tested for electrical safety at regular intervals by a competent person. The regulations also place a responsibility on the manufacturer of such appliances to ensure, as far as is reasonably practical, their products are safe when properly used.

A further responsibility is placed on service depots and repair workshops to ensure, on completion of a repair, that an item is electrically safe.

Where does the responsibility lie?

The responsibility starts with the owner of the venue. The fixed electrical installation should have been initially undertaken in accordance with the Institution of Electrical Engineers Wiring Regulations, 16th Edition, which whilst not statutory are widely recognised as a code of good practice. The primary means of protection against electrical shock is by adequate insulation and bonding of all external metal parts to earth. It is further recommended that all sockets to be used by entertainers are protected by RCDs, as discussed in chapter 7.

The venue owner is responsible for checking that any equipment brought into the venue conforms to the regulations. It is their duty to ask to see the entertainer's certificate of electrical safety.

Your responsibility

You must arrange for the **Portable Appliance Testing** of all your equipment at regular intervals, and carry a certificate to that effect. Each item will be marked as checked. If you do not bother you run the risk of having your performance stopped, or being barred from the venue. Being thus prevented from performing means not only that the Health and Safety Executive can instigate legal proceedings against you, but that your client could sue for breach of contract. Sounds a little drastic? Not at all. To my certain knowledge it has happened to a luckless few already.

Testing

Testing must be conducted by a suitably qualified electrical engineer, one who has had the necessary training to not only understand the procedure, but who can also interpret the result.

The tests begin with a **visual check** of the fuse rating and type, the wiring insulation and connections, and the case the item is housed in. It then moves on to **earthing tests.**

Insulation testing is also mandatory on all equipment. On **Class 1** equipment, that which is provided with an earth wire or earth point, 2 Mohms is acceptable. Equipment which does not require an earth is **Class 2**, and for this 7 Mohms is the norm.

There are of course, several other tests to be conducted.

Equipment register

When all items have been tested each has to be allocated a unique serial number. A list of those items must next be compiled in the form of a register, showing the results of the various tests, pass *and* fail.

Testing frequency

The average time between tests should be approximately three months, but dependent upon the condition of the equipment and frequency of use this period may be increased or decreased at the discretion of the tester.

Appendix 2

PORTABLE APPLICANCE TEST SHEET

APPLIANCE: MAKE: SERIAL NO: LOCATION: DATE OF FIRST TEST

TEST DATE / NAME OF TESTER (PRINT)	CONDITION OF FLEXIBLE CORD INCLUDING CORD GRIPS			MAINS ON/OFF SWITCH		INSULATION				EARTH CONT Ω	ACCESSIBLE FUSE HOLDER	EXPOSED OUTPUT CONNECTION	OVERALL PASS FAIL	COMMENTS	INITIALS
	VISUAL	POLARITY	PLUG*	VISUAL	OPERATION	CLASS	VISUAL**	INS RES M Ω	FLASH TEST		VISUAL NO LIVE PARTS EXPOSED	VISUAL			
/ /19	PASS FAIL	PASS FAIL	PASS FAIL	PASS FAIL	PASS FAIL	1 2	PASS FAIL				PASS FAIL	PASS FAIL			
/ /19	PASS FAIL	PASS FAIL	PASS FAIL	PASS FAIL	PASS FAIL	1 2	PASS FAIL				PASS FAIL	PASS FAIL			
/ /19	PASS FAIL	PASS FAIL	PASS FAIL	PASS FAIL	PASS FAIL	1 2	PASS FAIL				PASS FAIL	PASS FAIL			
/ /19	PASS FAIL	PASS FAIL	PASS FAIL	PASS FAIL	PASS FAIL	1 2	PASS FAIL				PASS FAIL	PASS FAIL			
/ /19	PASS FAIL	PASS FAIL	PASS FAIL	PASS FAIL	PASS FAIL	1 2	PASS FAIL				PASS FAIL	PASS FAIL			
/ /19	PASS FAIL	PASS FAIL	PASS FAIL	PASS FAIL	PASS FAIL	1 2	PASS FAIL				PASS FAIL	PASS FAIL			

* Verify Correct Fuse ** If Class 2, ⧈ is visible. † If voltage greater than 50v, test short circuit current

Fig. 16. Page from a PAT register.

Helping yourself

There is much you can do to help the tester in his job which will result in the test period being extended and which will, in turn, save you money.

- Replace any damaged plugs and sockets and make sure they contain the correctly rated fuse.
- Make sure all cables are effectively secured by the cable clamp.
- Check the casing of equipment for damage and loose/missing screws.
- Renew damaged or badly scuffed leads.
- Check for signs of over-heating.
- Has the item been used in conditions it is not suitable for? (ie excessive humidity and damp).

Contacts

For further information on Portable Appliance Testing you should consult your local Health and Safety Executive office whose address you will find in your telephone directory.

CHECKLIST

- Are you strictly legal and carrying the necessary licences? Do not be tempted to overlook this aspect of the job, especially PAT testing. An inspector is only likely to turn up when you are in the middle of a performance, unannounced, and is empowered to shut you down immediately. He will not be lenient and wait until the end of the evening.
- Make sure your paperwork is in order, and if you are unsure of anything ring the appropriate organisation and discuss your position. It will be dealt with sympathetically. These people only ever flex their muscles if you blatantly flout the rules and regulations.

DISCUSSION POINTS

1. Do you understand the difference between PRS, PPL, MCPS and VPL? What is the role of each?

2. Why is licensing so necessary?

3. What is the importance of PAT testing?

4. Are you familiar with the consequences of non-compliance of any of the above?

12
The DJ Associations

Like-minded tradesmen have always banded together to form associations, originally known as guilds. It is a practice which continues to this day. The voice of the tradesman can be better delivered collectively through an association, than by an individual alone. In some ways an association is similar to a trades union which exists to promote and protect the welfare, interests and rights of its members, mainly by collective bargaining. An association is only representative of its members, and is less militant in its approach. DJs are tradesmen and should, therefore, consider membership of an association to be an essential part of their organisation.

LOOKING AT THE BENEFITS

What's in it for me?

This is generally the first question people ask when approached to join a DJ association. Why are we so mistrustful? DJs are often accused of being *prima donnas* whose only interest is to feather their own nest and get as much as they can for themselves. This may well be so when some start their new career; they are probably very proud of their new disco and want to show it off as much as possible. Thankfully this is something which very soon passes, and they settle down to working and co-operating with fellow DJs quite amicably.

Am I good enough?

This is another consideration made by many which again is totally groundless. The typical DJ association positively welcomes everyone who is serious about their profession no matter whether they are a newcomer or seasoned veteran. Within the association the

Thames Valley Disc Jockeys Association
DISCOTHEQUE BOOKING AGREEMENT

An agreement made this day (as date below) between .... hereinafter called

the PROMOTER, and.. hereinafter called the DISCOTHEQUE, under which the PROMOTER engages the DISCOTHEQUE, and the DISCOTHEQUE accepts the engagement to appear at the venue(s), at the salaries, and on the date(s) shown on this agreement, and subject to the terms and conditions as printed on the reverse of one copy.

1. Address of PROMOTER ....

   ....Telephone number ....

2. Address of DISCOTHEQUE ....

   ....Telephone number ....

3. Venue and address ....

   ....Telephone number ....

4. Date(s) required ....Occasion ....

5. Performance times: FROM .... TO ....

6. The DISCOTHEQUE will arrive by .... for setting up of the equipment, and will vacate the premises by

   .... after dismantling and removing the equipment.

7. Approximate number of audience/guests .... Approximate age group of audience/guests ....

8. Type of music required ....

9. Agreed fee payable for the time of the performance as stated in (5) above .... if the PROMOTER requests that the

   DISCOTHEQUE extends the performance time, an additional rate of .... per hour will apply.

10. A booking deposit of .... must be paid with the return of the signed top copy of this agreement.

11. The balance of the fee of .... plus any additional fees, as stated in (9) above to be paid on the performance day.

12. The DISCOTHEQUE will provide:

    delete where inapplicable.

    adequate recorded music to suit the occasion .... YES / NO
    adequate equipment for playing and amplifying of recorded music .... YES / NO
    lighting effects to suit the occasion and venue .... YES / NO
    special effects or other services (by prior arrangement)............................ YES / NO
    disc jockey(s) .... YES / NO
    equipment operator/sound or lighting engineer(s).................................... YES / NO

13. The DISCOTHEQUE will comprise of: the discotheque equipment, disc jockey, and ....assistants, to allow setting up, operating, and dismantling the equipment in the prescribed times (5 * 6 * 9) above.

14. Any other special conditions ....

    ....

The TOP COPY of this agreement, duly signed must be returned within fourteen days of receipt, along with a booking deposit as stated in (10) above. All personal cheques must be supported with a current bankers guarantee card.

Please make cheques / money orders payable to ....

I THE UNDERSIGNED ACKNOWLEDGE THAT I HAVE READ THE ABOVE DETAILS, AND AGREE WITH THEM, AND HAVE ALSO READ THE TERMS AND CONDITIONS AS PRINTED ON THE REVERSE OF THE COPY OF THIS AGREEMENT, AND IN SIGNING, AGREE TO ADHERE TO THEM.

Signed, for and on behalf of the PROMOTER | Signed, for and on behalf of the DISCOTHEQUE

.... | ....

Promoter: Please retain the copy of this agreement for your reference. | date..

Fig. 17. Sample contract.

newcomer will learn much, sometimes with hands-on experience. One of the joys of being a DJ is that you never stop learning. New avenues are opening up all the time, and you will not learn much unless you mix socially with your contemporaries.

It pays to belong

The DJ associations are a voice to be heard within the disco industry and a force to be reckoned with. Since their inception they have between them been responsible for various items of legislation and codes of practice.

There are many benefits to be derived from membership.

Public liability insurance

Your membership fee will include an annual premium for public liability insurance, a necessary requirement. For any DJ to work without such cover is like a game of Russian roulette.

Picture the scene. Come Saturday night at a wedding reception, the best man offers to help by moving a heavy speaker. He accidentally drops it on his foot, and there in the post on Monday morning is a hefty writ suing you for damages, incapacity and loss of earnings. Without PLI cover where do you stand?

Using contracts

Another requirement of membership of most DJ associations is that you use a legally binding **contract** for all bookings. A scrap of paper with scribbled details is definitely no substitute, as you will soon discover should a customer refuse to pay, or welch on an agreement. By using a legally binding contract you could start legal proceedings to recover your fee. Such documents are generally available through the associations, and will have been checked legally as well as being approved by the Office of Fair Trading.

LEARNING FROM OTHER MEMBERS

Fees

One area of a DJ's work over which the associations have no control nor say is that of fees. Only an individual knows his own true worth and has the ability to negotiate a fair fee. By talking to other members, however, you can assess the local average, and by comparing what others have to offer you can decide on your own fee.

VIEWING MEMBERSHIP AS A QUALIFICATION

A source of bookings

DJ associations do not act as agencies in getting work for their members, but their office telephone number gets around and many potential customers see membership as a qualification. They book the disco with a degree of confidence.

Another way in which membership can give you extra bookings is by fellow members passing on their double bookings. Even if they have never seen you work, or the equipment you use, they will have confidence in you as you have taken the trouble to join the association. As with the agencies (see chapter 3), send in a date sheet to the association office, showing your availability, on a regular basis.

FIRE DRILL

Whilst it is not a condition of membership, most DJ associations recommend that members carry a suitable fire extinguisher. Most, if not all, venues have a selection readily available for use, but for your own peace of mind you should also carry your own. Ideally you need two, one in your vehicle and one with your disco equipment, each readily available for use in an emergency. Ask yourself:

- 'If a fire breaks out, do I know where the nearest extinguisher is, and more importantly, do I know how to use it?'

In the event of a fire, when the building needs to be evacuated you will be in charge as you have the public address system. Only you can give clear instructions for an orderly evacuation. Make sure you know where the emergency exits are, and more importantly, ensure they are free from obstructions. All too often mobile discos are told to set up in front of emergency exits. This is illegal, so politely refuse, pointing out the error of the organiser's request. Failure to do this makes you, not the person who told you where to set up, liable for the obstruction.

Fire extinguisher types and uses

There are basically five different types of extinguisher, each designed to deal with different sorts of fire. Get to know each. They are colour-coded for easy recognition.

The types of fire you are likely to encounter are:

- A—**solids**, such as wood, paper, textiles.
- B—**liquids**, such as petrol, oil, fat, paint, solvents.
- C—**gases**, such as propane, butane, natural gas.
- D—**electrical**, as in electrical equipment.

The five colour codes of these extinguishers are:

- **Red**—water or water/gas, for Group A.
- **Cream**—foam, for Groups A and B.
- **Black**—CO_2, for Groups B, C and D.
- **Blue**—dry powder, for Groups B, C and D.
- **Green**—BCF (halon), for Groups A, B, C, and D.

It should be noted that the green one is difficult, if not impossible, to obtain as it contains an ozone-depleting CFC gas. Foam is suitable for use on 12v DC electrical equipment and in vehicles.

ENJOYING THE SOCIAL BENEFITS

If for no other reason, you should find membership of an association beneficial in giving a tremendous boost to your social life. Meetings are held on a monthly basis in an informal way, which together with the newsletter that each sends out gives scope to make new friends with similar interests to your own. As far as is practical, some associations organise a series of special nights away from the disco environment. Thames Valley DJ Association hold an Awards Presentation Night complete with dinner at a top venue each year, and the South Eastern Discotheque Association holds an annual equipment exhibition which has risen in standing to become the largest regional show of its kind in the UK. Most of the associations are regularly invited to the factories where equipment is made, to shop openings, and get special concessionary prices on many things.

Contacts

Central Association of DJs: Carole Lowe. Tel. (0121) 477 2757.
Discos In Shire Counties Organisation: Sue Mansell. Tel. (01827) 251193.
South Eastern Discotheque Association: Roger Eagleton. Tel. (01795) 890436.
Thames Valley DJ Association: Mike Jordan. Tel. (01734) 771450.

There is no central body overseeing DJ associations as there is with, for example, hospital radio stations.

CHECKLIST

- No matter what your status as a DJ, be it in mobile disco, pubs and clubs, radio, or even bedroom mixing, you should consider joining a DJ association where you will not only gain useful information, assistance and new friends, but will be put into a stronger position should you need to voice your views on a topic.

- As a member you will be extended certain concessions when it comes to purchasing new equipment and gaining admission to certain events.

- You become part of a closely knit family who are all keen to help each other.

CASE STUDIES

Henry joins a DJ association

Henry, aged 21, and his friend decide to start a mobile disco. There is a retailer selling the necessary equipment in town, but neither Henry nor his friend know where to start in selecting their items, and do not wish to look stupid by asking. They had a look around the shop, but came away feeling more confused than ever. They decide that perhaps before buying anything they should read a few books on the subject to get a rough idea of what to look for. The only books the library can offer are highly technical. They really need something more basic.

When Henry's friend was browsing through adverts for mobile discos in *Yellow Pages* he noticed several mentioned membership of an association. Henry rings one and asks for a contact phone number. The DJ gladly gives it, asks if Henry is interested in learning more and offers to take him along to a meeting the following weekend. Henry accepts the invitation, and after going to several meetings begins to piece together the answers to his earlier queries. He also sees that the association runs a 'shop', selling various disco items at cost price.

George makes a good move

George, aged 25, has recently moved into the town, having been transferred from a nightclub some distance away. He knows no one in town, and has little time for socialising and making new friends due to the unsocial hours of his job. George is a dedicated DJ and decides to make contact with the nearest association. Here he starts to make new friends, help newcomers with his experience as a nightclub DJ, and arranges for members to visit his club at reduced admission prices whenever they wish.

DISCUSSION POINTS

1. What is the significance and purpose of a DJ association?

2. How does its role differ from that of a regular trades union?

3. List the pros and cons of membership which you can see.

4. Which of the pros and cons outweigh each other, and why do you think this is so?

5. How does membership of a DJ association affect your success or failure as a DJ?

Glossary

ac. Alternating current.

AM. Amplitude modulation. The radio signal formerly known as medium wave.

Bass reflex. A type of speaker which has a port, or hole, in the baffle to increase its bass response.

Bias. The natural inward pull of a moving pick-up arm.

Cans. Slang for headphones.

Cardioid. A type of microphone with a heart-shaped area of sensitivity in a forward direction giving high gain from the front, but reduced gain from the sides and rear.

Cartridge. The end of a pick-up arm which converts the signal from the stylus into electrical energy before reaching the pre-amp. It is also the name given to a cassette of endless loop tape used by radio stations for jingles and so forth. In this instance it is universally abbreviated to 'cart'.

CD. Compact disc. The new generation for reproducing music which is taking over from vinyl records. It uses digital technology, as opposed to analogue, to offer near-perfect reproduction.

Cone. Specially treated paper, card, or plastic used as a moving speaker component.

Crossover network. Circuitry which controls frequency ranges feeding different loudspeakers.

Crystal microphone. A type of microphone in which a fluctuating voltage is generated by applying pressure to a material such as rochelle salt or barium titanate. Usually a cheap microphone not really worth considering for serious work.

Cue/cue up. To bring a record, tape or CD to the desired starting point ready to play.

DAT. Digital audio tape. A compact tape recording medium which closely resembles a dictaphone cassette.

dc. Direct current.

Demo tape. Demonstration tape. A tape compiled by a broadcaster to demonstrate his/her skills.

Dubbing. Transferring a recorded sound from one recording to another. It can also mean mixing more sound on to a recording.

Dynamic microphone. A microphone which operates through a moving coil.

DBX. A noise reduction system.

db. Decibel. One-tenth of a bel, a unit of measurement for sound.

Digital. In numerical form, CDs, samplers and DATs all store music digitally, as opposed to cassettes and records which are analogue and store their sounds electrically. Digital storage eliminates surface noise problems and gives near-perfect sound reproduction.

DIN. Deutscher Industrie Normanausschuss. A German standard of measurement widely used throughout Europe for equipment and performance.

DNL. Circuitry designed by Philips incorporated into tape and cassette recorder playback amplifiers to reduce tape noise.

Dolby. Circuitry designed for the reduction of background noise on recorded tapes, audio or video. A system developed in the Dolby Laboratories Inc.

Dropout. An audible defect on a tape.

Dual concentric. A loudspeaker system in which a high frequency cone is fitted within a low frequency one.

Electret. A type of microphone which needs a permanent power supply such as a battery to operate it.

Electrostatic microphone. Sometimes also referred to as a **condenser microphone**. A type in which a fluctuating electric current is produced by the movement of a diaphragm relative to a rigid plate.

Elliptical stylus. A stylus manufactured with an elliptical-shaped tip and with an oval contact pattern.

Equalisation. Circuitry which modifies frequency response throughout the full audio frequency range commonly performed by adjusting treble and bass controls.

FM. Frequency modulation. The radio signal formerly known as VHF.

Feedback. When the output from a speaker is picked up by a microphone and fed back through the amplifier to the speaker again and again in a circle, resulting in a high-pitched squeal.

Flutter. Rapid fluctuations of a tape or record speed.

Foldback. Stage monitor speakers.

Format. The definition of a particular radio station's type of programming.

Frequency. The number of sound wave peaks which are formed at a given point over a period of time. The greater the number, the higher the frequency and the lower the wavelength.

Gaffa tape. Material-backed tape used for sticking down cables and leads.

Graphic equaliser. A device which enables selective control of the different frequencies. Each frequency band is controlled by a separate fader.

Hemispherical stylus. A stylus manufactured with a circular-shaped tip.

IEC. International Electrotechnical Commission. The standard for electric and electronic equipment.

ILR. Independent local radio.

INR. Independent national radio.

Impedance. The natural opposition to an ac current which is measured in ohms.

Intro. The musical instrumental introduction of a song which precedes the vocal section.

IRN. Independent Radio News, the company which supplies ILR stations with their national news output.

Jack plug. A two- or three-contact audio plug based on an old Post Office design. The most widely used size is the .625mm or 1/4in.

Jingle. A short radio station or programme identification message. Used to also be widely favoured by disco DJs. Usually found on carts.

Jock. A shortened version of disc jockey.

Joule. A measurement of energy usually referred to in connection with strobe lighting.

KHz. Kilohertz. One thousand hertz, or cycles, the hertz being a unit of radio frequency named after the 19th-century physicist Heinrich Hertz who was the first person to produce radio waves artificially.

Leader. Red and green open reel tape used to mark the beginning and ending of a track or section on a tape.

LED. Light emitting diode.

Links. An announcement connected to the previous item in a programme or one which introduces the next.

MHz. Megahertz. One million hertz, or cycles.

Midi. The communications standard that allows most items of audio equipment to communicate with each other. It is a digital signal carried along 3-pin DIN cables.

Mid-range. A type of loudspeaker which utilises the audio frequencies of 300–3,000 Hz.

Mix. A succession of music with a matching beat or rhythm.

Mixer. A unit which allows the outputs of several channels to be individually faded or mixed.

MOSFET. Metal oxide semi-conductor field effect transistor. An amplifier output device.

NAB. National Association of Broadcasters.

OB. Outside broadcast. A radio programme which is transmitted from a location outside the confines of the studio.

Ohm. A unit to measure resistance or impedance.

On-air. The physical act of broadcasting live.

Omnidirectional microphone. A microphone with an equal sensitivity to sound from all around it.

Padding. The art of filling in the spare seconds when a machine malfunctions or you have mistimed something.

Patchbay. A rack-mounted collection of sockets which allows equipment to be interconnected.

PPM. Peak programme meter. Similar to a standard VU meter found on most disco consols, but a PPM allows for a faster attack time and a slower decay. More widely found in radio studios.

Phono plug. A type of small coaxial plug often found on domestic hi-fi equipment.

Piezo horn. A treble loudspeaker horn with no moving parts.

Potentiometer or Pot'meter. A volume or tone control.

Pre-amp. A small, low-powered amplifier used to increase the signals from microphones, turntables and so forth to a sufficient level to be amplified by a power amp which feeds the loudspeakers.

Prefade listening or PFL. A method of listening to a certain mixer channel before it is faded up.

RSL. Restricted service licence. A licence awarded to community radio groups to broadcast experimentally for up to twenty-eight days at a time.

Rumble. Unwanted turntable or motor noises amplified through a sound system.

Sampler. An electronic tape recorder that can reproduce sounds on cue.

Segue. A term used to describe a succession of uninterupted music.

Splice. Cutting and joining a tape in the process of editing.

TBU. Telephone balancing unit. A studio telephone switchboard.

Triac. A switching device used in sound-to-light controllers.

Tweeter. A high-frequency range of speaker.

Voice-over. The spoken narrative of a documentary, commercial, or radio station ident over a music bed.

Vox pop. An unrehearsed, on-the-spot interview with a member of the public. The passer-by is usually asked for his/her opinions/views on a topic, then later all the answers are spliced together, cutting out the question.

VU meter. Found on typical disco consols to measure audio levels in volume units.

Watt. A unit used to express electrical and acoustic power.

Woofer. A low-frequency range of speaker.

Wow. Slow fluctuations of tape or record speed.

Appendix
Lighting and Audio Plugs

BULGIN LIGHTING PLUG

These are the industry standard plugs and sockets for connecting lighting units. Three or four channels can be wired using the same plugs. There is no set colour coding for wiring them so you must ensure that the coloured wires in a plug correspond with those in the socket. Pin number 1 is always used for earth. If wiring for a three-channel system use pins numbers 3, 4 and 5 to take the live feed, pins numbers 6, 7 and 8 the neutral. If your system is of four channels, pin number 2 should also be connected.

When wiring up for a three-channel system, for example, you can use five-core cable. Then instead of connecting pins numbers 6, 7 and 8 individually, one wire can be used for the three, as shown in Fig. 3.

1/4IN AUDIO JACK PLUG

Most mobile discos use 1/4in jack plugs and sockets on their speaker systems. The larger shows, using more powerful and expensive equipment, use Cannon XLR connectors. All audio leads should be screened, ie using cables with braiding material around the inner insulated cables. This type of cable is safer to use and will give fewer audio problems.

Cut away the plastic covering from the inner cable by 1/4in. Twist the strands of wire together and coat with solder (tinning). Solder this to the short tab. If it is a stereo lead you are putting together, it should be similarly dealt with and soldered to the other tabs. The screening braid should also be connected to these tabs. Make sure when finished that the cables or the solder are unable to touch the outer cover when threaded on.

PHONO AUDIO PLUG

These plugs are more widely used in domestic hi-fi units, but are used in disco systems to connect turntables, CD players, cassette players and so forth to the mixer. They should be wired up in much the same way as a 1/4in jack plug.

5-PIN DIN AUDIO PLUG

There is no standard when wiring up 5-pin din plugs/sockets. They have been used for so long, principally for domestic hi-fi equipment, that different manufacturers settled upon their own individual wiring combinations. The most common are:

- Pins numbers 1 and 4 are the inputs.
- Pins numbers 3 and 5 are the outputs.
- The red wire is usually soldered to pins numbers 1 and 4.
- The yellow wire is usually soldered to pins numbers 3 and 5.
- Pin number 2 is the earth.

Useful Addresses

SETTING UP THE BUSINESS

Companies House, Crown Way, Cardiff CF4 3UZ. Tel: (01222) 388588. To register your name.

Atlantic Print, 14 Oxen Lease, Singleton, Ashford, Kent TN23 2YT. Tel: (01233) 624538. Printing.

MUSIC

Discomix Club, PO Box 89, Slough, Berks SL1 8NA. Tel: (01628) 667276.

Music Factory, 5/7 Fitzwilliam St, Parkgate, Rotherham, South Yorks S62 6EP. Tel: (01709) 710022.

Trax, 82 Station Road, Birchington, Kent. Tel: (01843) 848494.

LICENSING

ASP Frequency Management Ltd, Edgcott House, Lawn Hill, Aylesbury, Bucks HP18 0QW. Tel: (01296) 770458. Radio microphones.

Performing Rights Society, 29/33 Berners Street, London W1P 4AA. Tel: (0171) 580 5544.

Mechanical Copyright Protection Society, Elgar House, 41 Streatham High Road, London SW16 1ER.

Phonographic Performance Ltd, Ganton House, 14/22 Ganton Street, London W1V 1LB. Tel: (0171) 437 0311.

ASSOCIATIONS

National Association of Hospital Broadcasting Organisations,

Milne House, 1 Norfolk Square, London W2 1RU. Tel: (0171) 402 8815.

Student Radio Association. Tel: (01602) 513617.

Community Radio Association, The Media Centre, 15 Paternoster Row, Sheffield S1 2BX. Tel: (0114) 279 5219.

Community Radio Association (London Development Unit), Lambeth College, Belmore Street, London SW8 2JY. Tel: (0171) 738 8788.

Central Association of DJs. Tel: (0121) 477 2757.

Discos in Shire Counties Organisation. Tel: (01827) 251193.

South Eastern Discotheque Association. Tel: (01795) 890436.

Thames Valley DJ Association. Tel: (01734) 771450.

The Agents Association, 54 Keyes House, Dolphin Square, London SW1V 3NA. Tel: (0171) 834 0515.

Further Reading

BOOKS

Electrical Safety for Entertainers (Health and Safety Executive).

The Guinness Book of British Hit Albums, Paul Gambaccini, Tim Rice and Jonathon Rice (Guinness Publishing).

The Guinness Book of British Hit Singles, Paul Gambaccini, Tim Rice and Jonathan Rice (Guinness Publishing).

The Guinness Book of Top Forty Charts, Paul Gambaccini, Tim Rice and Jonathan Rice (Guinness Publishing).

How to Do Your Own Advertising, Michael Bennie (How To Books Ltd, 1996).

How to Get into Radio, Bernie Simmons (How To Books Ltd, 1995).

How to Manage Budgets and Cash Flows, Peter Taylor (How To Books Ltd, 1994).

How to Market Yourself, Ian Phillipson (How To Books Ltd, 1995).

Making It as a Radio or TV Presenter, Peter Baker (Piatkus Books, 1995).

The Omnibus Chart Book of the 80s, Dave McAleer (Omnibus Press).

The Radio Authority Pocket Book, (The Radio Authority Press and Information Office).

Take to the Air, Theo Short and Richard Waghorn (Services to Business Dept, Trinity and All Saints College, Leeds, 1995).

The Radio Handbook, Pete Willy and Andy Conroy (Routledge, 1995).

How to Set Up a Home Recording Studio, (PC Publishing, 4 Brook Street, Tonbridge, Kent TNP 2PJ, 1995).

MAGAZINES

Disco Mirror, Waterloo Place, Watson Square, Stockport SK1 3AZ. Tel: (0161) 429 7803.

DJ Magazine, 4th Floor, Centro House, Mandela Street, London NW1 0DU. Tel: (0171) 387 3848.

Lighting and Sound International, 7 Highlight House, St Leonards Road, Eastbourne, Sussex BN21 3UH. Tel: (01323) 642639.

Mix Mag, DMC Ltd, PO Box 89, Slough, Berks SL1 8NA. Tel: (01628) 667124.

Mixology, Music Factory Ltd, 29 Alfred Street, Kettering, Northants NN16 0SW. Tel: (01536) 522267.

Mobile DJ, PO Box 93, Grimsby, Lincs DN36 4GF. Tel: (01472) 827311.

Index

HOW TO GET INTO RADIO
Starting your career as a radio broadcaster

Bernie Simmons

We are about to see a huge expansion in radio stations, and digital audio broadcasting is set to revolutionise the industry. Radio broadcasting now offers an established career path, with industry-approved qualifications like NVQs and university degrees in radio broadcasting. There has never been a more exciting time to make a career in radio. But how do you get in? Where do you get started? What training is on offer? All this and more is revealed in this readable and up-to-the-minute new book. 'Useful to any aspiring jock—contains a lot of good advice.' *The Stage & Television Today*. Bernie Simmons is a professional radio broadcaster with a wealth of varied experience. He himself started out in night club DJ-ing, hospital radio, and in-store radio. He has since worked on breakfast shows on independent local radio, news magazines and phone-in programmes on community radio, worldwide radio services such as BFBS Radio and The BBC World Service, Gold AM Radio and the pioneering Satellite Radio.

153pp. illus. 1 85703 143 1.

HOW TO DO YOUR OWN ADVERTISING
The secrets of successful sales promotion

Michael Bennie

'Entrepreneurs and small businesses are flooding the market with new products and services; the only way to beat the competition is successful selling—and that means advertising.' But what can you afford? This book is for anyone who needs— or wants—to advertise effectively, but does not want to pay agency rates. Micheal Bennie is Director of Studies at the Copywriting School. 'An absolute must for everyone running their own small business . . . Essential reading . . . Here at last is a practical accessible handbook which will make sure your product or service gets the publicity it deserves.' *Great Ideas Newsletter* (*Business Innovations Research*). 'Explains how to put together a simple yet successful advertisement or brochure with the minimum of outside help . . . amply filled with examples and case studies.' *First Voice* (National Federation of Self Employed and Small Businesses).

176pp. illus. 1 85703 213 6. 2nd edition.

HOW TO WRITE FOR TELEVISION
A complete guide to writing and marketing TV scripts

William Smethurst

Television is a huge and expanding market for the freelance writer. Particularly in the field of drama, producers are constantly looking for new writers for situation comedies, series drama, and soap operas and single plays. But what kind of scripts are required? How should a script be presented and laid out? What camera moves should you put in, and should you plan for commercial breaks? What programmes and organisations should you contact, and which are the subjects to tackle or avoid? Packed with hard-hitting information and advice, and illustrated throughout with examples, this is a complete step-by-step manual for every writer wanting to break into this lucrative market. 'Packed with information which is well presented and easily accessible.' *National Association of Careers & Guidance Teachers Bulletin.* 'If would be TV scriptwriters are looking for a wide ranging and practical book to light the fuse which could lead to a successful career, they should certainly invest in a copy of William Smethurst's *How to Write for Television.*' *BAFTA News.* 'Your best starting point is probably William Smethurst's book.' *Writers News.* William Smethurst has been a television script editor at Pebble Mill, and executive producer of drama serials for Central Television. He is now a director of the Independent television company, Andromeda Television Ltd.

160pp. illus. 1 85703 045 1.

HOW TO MARKET YOURSELF
A practical guide to winning at work

Ian Phillipson

In today's intensely competitive workplace it has become ever vital to market yourself effectively, whether as a first-time job hunter, exisiting employee, or mature returner. This hard-hitting new manual provides a really positive step-by-step guide to asserting yourself, choosing the right personal image, identifying and presenting the right skills, building confidence, marketing yourself in person and on paper, organising your self-marketing campaign, using mentors at work, selling yourself to colleagues, clients and customers, and marketing yourself for a fast-changing future. The book is complete with assignments and case studies.

160pp. illus. 1 85703 160 1.

HOW TO GET INTO FILMS AND TV
Foreword by Sir David Puttnam

Robert Angell

Would you like to make a career in films or television? Whether you want to direct feature films, photograph documentaries, edit commercials or pop videos, write current affairs programmes for television, do art work for animation or just know that you want to be involved in film or television in some capacity but are not quite sure how to set about getting started, this book will give you a wealth of information to guide you through the dense but exotic jungle of these exciting industries. 'Readable and useful.' *Amateur Film and Video Maker*. 'An indepth coverage of the subject. Offers a wealth of useful advice and addresses for more information . . . One of the essential references for careers libraries.' *The Careers Officer Journal*. 'A comprehensive guide in lay language . . . Each section includes suggested starting points for newcomers.' *BAFTA News*. 'You'll find all the answers to your questions in Robert Angell's book.' *Film Review*. Robert Angell is a Council Member of the British Academy of Film & Television Arts (BAFTA) and Chairman of its Programme Committee and Short Film Award jury.

144pp. illus. 1 85703 162 8. 3rd edition.

HOW TO BE A FREELANCE JOURNALIST
Your step-by-step guide to success

Christine Hall

Writing articles and features is a skill which can be learned by everyone with average ability and intelligence. Based on firsthand experience, this new book shows step-by-step how you can break into print, how to develop your ideas into publishable articles, and how to sell them, and how to develop a profitable hobby into a fulltime freelance career. Written from the editor's viewpoint, this is an invaluable source of insider knowledge for writers at all levels. Christine Hall has wide experience as a subeditor, production editor, features editor, copy editor, deputy and acting editor for newspapers, consumer and trade magazines in the UK, Germany and China. A Member of the Society of Authors, and of the Society of Women Writers and Journalists, she has taught the craft of writing to beginners and advanced students alike.

160pp. illus. 1 85703 147 4.

HOW TO DO YOUR OWN P.R.
Getting the right publicity for your organisation

Ian Phillipson

Public relations is one of the best ways of promoting any organisation, business or event. It is not as costly as advertising, and properly used can produce far better results. This new book provides a really practical step-by-step guide to doing your own P.R., whether you want a one-off press release, or a more formal and wide ranging campaign. It shows how to recognise and create stories that interest the media, how to write and lay out press releases, how to be interviewed by the press, as well as providing a bank of proven ideas for public relations stories. The book is complete with helpful checklists and realistic case studies. Ian Phillipson is an experienced public relations consultant with a range of clients in the public and private sectors.

128pp. illus. 1 85703 145 8.

CREATIVE WRITING
How to develop your writing skills for successful fiction and non-fiction work

Adele Ramet

The term 'Creative Writing' covers a broad spectrum of skills from writing non-fiction articles and features for specialist magazines to romantic fiction, ghost stories and crime novels. This book guides you through key techniques, with exercises aimed at helping you to write more effectively in your chosen genre. You will be encouraged to approach your work from different angles, demonstrating how to bring a fresh slant to non-fiction pieces and how to involve yourself more fully in the lives of your fictional characters. Whatever your writing interest, this book will help you write more creatively and lead you further along the route towards publication. Adele Ramet is Chairman of the South Eastern Writers Association and an experienced writing tutor. She has contributed widely to *Bella, Woman's Realm*, and many other leading women's magazines.

144pp. illus. 1 85703 451 1.